BIOGENIC AMINES
AND PHYSIOLOGICAL MEMBRANES
IN DRUG THERAPY
(in two parts)

Part A

MEDICINAL RESEARCH:
A SERIES OF MONOGRAPHS

Consulting Editor: GARY L. GRUNEWALD

VOLUME 1: *Drugs Affecting the Peripheral Nervous System, edited by Alfred Burger*
VOLUME 2: *Drugs Affecting the Central Nervous System, edited by Alfred Burger*
VOLUME 3: *Selected Pharmacological Testing Methods, edited by Alfred Burger*
VOLUME 4: *Analytical Metabolic Chemistry of Drugs, by Jean L. Hirtz*
VOLUME 5: *Biogenic Amines and Physiological Membranes in Drug Therapy,*
Edited by John H. Biel and Leo G. Abood

In Preparation:

VOLUME 6: *Search for New Drugs, Edited by Alan D. Rubin*

BIOGENIC AMINES AND PHYSIOLOGICAL MEMBRANES IN DRUG THERAPY

(in two parts)

PART A

Edited by

JOHN H. BIEL

DIRECTOR, DIVISION OF PHARMACOLOGY AND MEDICINAL CHEMISTRY
ABBOTT LABORATORIES, SCIENTIFIC DIVISIONS
NORTH CHICAGO, ILLINOIS

and

LEO G. ABOOD

CENTER FOR BRAIN RESEARCH AND DEPARTMENT OF BIOCHEMISTRY
UNIVERSITY OF ROCHESTER
ROCHESTER, NEW YORK

1971

MARCEL DEKKER, INC., NEW YORK

PREFACE

Modern methods in biochemistry, electrophysiology and histology have provided valuable tools in the study of neurotransmission, the elucidation of the mechanism of action of psychotropic and cardiovascular drugs and the biochemical triggering or mediation of disease processes.

These developments have invoked a major role for the biogenic amines and the cell membrane as important biological factors to be "manipulated" with respect to the chemical therapy in at least three major disease areas: mental as well as certain neurological disturbances, cardiovascular illness, and allergic responses.

During the last three years an increasing amount of experimental evidence has appeared to indicate that the action of the psychotropic and cardiovascular drugs is not merely a consequence of an alteration in the balance of certain endogenous amines, but that the cell membrane plays an intimate part in bringing about the characteristic effects induced by a widely varying structural array of CNS and cardiovascular agents.

To gain some insight into the possible mode or modes of action of experimental as well as established drugs which affect the central and peripheral nervous system, the editors felt that three major topics should be included in a treatise of this type:

1. Biogenic amines thought to be implicated in neurotransmission and the mechanism of action of psychotropic, cardiovascular and anti-allergic drugs.

2. The nature of the cell membrane, particularly as it relates to the transport of biogenic amines and drugs.

3. The subtle role of the chemical structure of a drug with respect to the nuances in the biological responses it evokes.

The book is designed to provide both a theoretical as well as pragmatic basis to an understanding of some major fields of modern drug therapy and should, therefore, be of immediate value not only to the basic scientist working in the general area of therapeutics, but also to graduate students

iii

in pharmacology, biochemistry, physiology and pathology, as well as senior medical students who should become increasingly concerned with the intricacies of current and future drug therapy.

It is hoped that this book will raise as many questions as it will answer to stimulate increased research pursuit in these areas and contribute to a more rational approach to the development of new drugs and their therapeutic application.

John H. Biel, PH. D.
Leo G. Abood, PH. D.

CONTRIBUTORS TO PART A

Leo G. Abood, Center for Brain Research and Department of Biochemistry, University of Rochester, Rochester, New York

Ross C. Bean, Philco-Ford Corporation, Aeronutronic Division, Newport Beach, California

Joellen T. Eichner, Philco-Ford Corporation, Aeronutronic Division, Newport Beach, California

Akira Matsubara, Center for Brain Research and Department of Biochemistry, University of Rochester, Rochester, New York

William C. Shepherd, Philco-Ford Corporation, Aeronutronic Division, Newport Beach, California

Ryo Tanaka, Center for Brain Research, University of Rochester, Rochester, New York

Ichiji Tasaki, Laboratory of Neurobiology, National Institute of Mental Health, Bethesda, Maryland

CONTENTS OF PART B

CONTENTS

BIOGENIC AMINES

AND PHYSIOLOGICAL MEMBRANES

IN DRUG THERAPY
(in two parts)

Part A

Chapter 1

THE ROLE OF PROTEINS AND LIPIDS IN
MEMBRANE STRUCTURE AND FUNCTION*

Leo G. Abood and Akira Matsubara†

Center for Brain Research and Department of Biochemistry
University of Rochester
Rochester, New York

*This research was supported by National Institutes of Health grants
NB-05856 and NB-06827.

†Present address: Laboratory of Chemistry, Department of General
Education, Kyushu University, Fukuoka, Japan.

1

I. INTRODUCTION

A fundamental assumption underlying the present discussion is that the action of many drugs on living systems is exerted on the membrane and the variety of processesit controls. If one examines the ultrastructural features of a cell, it becomes immediately evident that the cell is comprised of a large variety of membranous components and organelles, each in its turn made up of an intricate membranous network. Many important cellular functions such as energy synthesis, protein synthesis, transport, and nerve conduction occur within membranes. The term "membrane" is, however, a complex one from a biological standpoint, whereas, in each of the cellular functions alluded to, the particular role ascribed to the membrane is distinct. In the case of energy (i.e., mitochondrial ATP) synthesis, the membrane provides the matrix for the precise structural organization of enzymes and cofactors favorable for linking oxidation-reduction reactions to ATP synthesis. At the other extreme the excitatory (nerve) membrane is a dynamic macromolecular complex capable of undergoing rapid reversible changes in response to chemicals and ionic shifts, where enzymic reactions may not be directly involved. Despite the functional distinction between the mitochondrial and excitatory membrane, in both instances the membrane can be conceived as the structural matrix peculiar to function and comprised of some homologous chemical constituents. Although the primary focus of the present discussion is on neuronal membranes, particularly those components involved in excitatory activity, frequent reference is made to muscle membranes and other membrane components of the neuron.

II. MECHANISMS WHEREBY DRUGS INFLUENCE MEMBRANE PROPERTIES

There are a variety of ways in which drugs and other chemicals can influence membrane structure and function, among the more obvious of which are the following: (1) by modifying directly the activity of a membranous enzyme system; (2) by interaction with a membranous lipid in such a way as to influence the overall configuration of the membrane or the possible lipid-requirement of an enzyme; (3) by interaction with a structural protein so as to alter membrane structure or physical characteristics, such as viscosity, charge distribution, and stereospecificity; (4) by physical occlusion of membrane lattices or pores; (5) by altering the physical state of membranous water; (6) by altering the ion-exchange (permselective) characteristics of the membrane by affecting the overall charge distribution; (7) by obstructing, facilitating, or simulating the adsorption of some endogenous mobile cofactor for membrane structure or function; (8) by chemical combination with a membrane constituent (i.e., covalent bonding); (9) by interacting with the polysaccharide component of the membrane.

The present discussion in no way attempts to review the extensive literature in this area but merely to present some examples of drugs acting by such mechanisms. Its major objective is to describe the use of artificial membranes in approaching the problem of membrane physiology and to emphasize particularly the role of structural proteins.

III. SOME THEORIES OF MEMBRANE STRUCTURE

Although it is known that the synaptic and other membranous components of the neuron are comprised of proteins, lipids, and polysaccharides, there still exists considerable controversy regarding the interrelationship of the three and their relative importance in determining the basic configuration. Ever since the introduction of the Danielli-Davson model (1) of a membrane, the tendency has been, until recent years, to assign to lipids the primary role while proteins and polysaccharides assumed a secondary structural role. According to this model, lipids were arranged in the form of bimolecular parallel leaflets with the protein attached to the outer hydrophilic surface as an extended polypeptide chain [Fig. 1(b); proteins not included in models]. Modifications of this model have been proposed which arrange the lipids in the form of globular or elliptical micelles [Fig. 1(c)] or a combination of the parallel and globular structures [Fig. 1(d)]. It is likely that all arrangements are found depending on the membrane source. In myelin, for example, the most likely configuration is the bimolecular parallel array, whereas in more permeable membranes the arrangement of lipids may be globular or elliptical.

Recently, however, electron microscopists have proposed that the basic structural matrix of the membrane may be proteinaceous rather than lipoidal

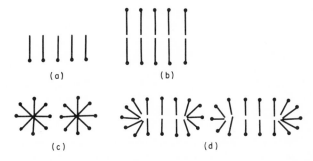

Fig. 1. Various molecular configurations for lipids. (a) parallel monomolecular array (e.g., condensed surface film); (b) parallel monomolecular array (e.g., bimolecular lipid film); (c) globular micelle (lipid dispersed in aqueous phase); (d) elliptical micelle comprised of both (b) and (c).

(2,3). The main evidence for this is that removal of the lipids at low
temperature under the mildest conditions does not appreciably change the
ultrastructural characteristics of the tissue preparations prior to lipid removal. Although such evidence is very convincing, it might still be argued
that the fixative procedures (e.g., glutaraldehyde), by denaturing and
dehydrating proteins, may impart certain ultrastructural characteristics
to the membrane; so that such stabilization is insufficient to prevent alteration after removal of the lipid. In any case it is apparent in the subsequent
discussion that the role of proteins in determining membrane structure and
function is an extremely important one, if not the major one.

One of the most easily prepared membrane fractions of mammalian
tissues, including brain, is the so-called "microsomal" fraction. The
microsomal fraction of beef brain is a heterogeneous conglomerate of
numerous membranous and other cytoplasmic components. Included among
these are fragments of synaptic endings, synaptic vesicles, endoplasmic
reticulum, axonal-dendritic processes, nuclei, and glial processes [Fig.
2(a)]. When an aqueous suspension of microsomes is sonicated, the
majority of the membranous components is fragmented to much smaller
macromolecular components which appear in the form of aggregates [Fig.
2(b)]. Many vesicular components (possibly synaptic vesicles) are still
present as well as small membrane fragments exhibiting a fine structure
not unlike that of the vesicles and other intact membrane fractions. The
small membranes appear to be comprised of globular arrays of electron-
dense particles.

It cannot be determined which of the membranous components are
residuals of the membranes existing prior to sonication and which are re-
formed from smaller components; however, even after ultracentrifugation
at 250,000 x g of the sonicated microsomal fraction, the ultrastructural
features of the supernatant material are the same as that of the material
prior to centrifugation. Evidently, many of the larger fragments resembling small nerve endings, vesicles, and myelin-like membranes could, as a
result of fixation, have re-formed from the smaller lipid-protein micelles
in the sonicated dispersion. It is not unlikely, therefore, that many of the
membranous components seen in isolated cytoplasmic fractions of brain
tissue do not represent membranes derived directly from cellular components. They may have been reconstitututed of macromolecular components
derived from more than one membranous source. For this reason caution
should be exercised in interpreting data on enzyme localization in isolated
cytoplasmic fractions, particularly where disintegration of the organelles
(e.g., mitochondria, nerve endings, and endoplasmic reticulum) may have
occurred.

One way of working with relatively pure membranous components is to isolate cellular organelles such as mitochondria, nerve endings, or myelin and then, by various fragmentation procedures and density gradient (4-6) centrifugation, to isolate the membranous components. Since nerve endings are abundant in brain tissue, can readily be isolated, and are important excitatory components, considerable attention has been directed to them as a source for a homogeneous membrane preparation. Nerve endings which are first isolated from rat brain by discontinuous Ficoll-density gradient centrifugation (4) are fragmented by hypotonic shock, and the fragmented preparation is then subjected to a discontinuous sucrose gradient to sep-arate the synaptic membranes (6,7). An electron micrograph of such a preparation is presented in Fig. 3. It should be emphasized, however, that in all subcellular fractions of brain both glial and neuronal components are present; and until such time as satisfactory methods are available for the separation of both cell types, this problem should be borne in mind.

IV. MEMBRANES PREPARED FROM LIPIDS AND PROTEINS

Another approach to the problem of membrane structure is to prepare artificial membranes from various structural components of biological membranes and to compare their fine structure to that of the intact mem-branes. Numerous investigators [see (8) for review] have examined the ultrastructural features of lipid membranes prepared in a variety of ways. An example of such a preparation (9) is presented in Fig. 4(a). Such lipid membranes, which required the presence of Ca^{2+} for their formation (9), assume a ribbon-like configuration having a thickness of about 100 Å and widths varying between 100-400 Å. High resolution electron microscopy (200,000 x) reveals that the ribbon-like structure actually consists of a double membrane, possibly a bimolecular parallel arrangement of the lipid molecules.

Occasionally, laminated structures are found which bear a resemblance to myelin. It is of interest that Ca^{2+} or another divalent cation is required for the formation of the bimolecular configuration seen in the micrographs, while in its absence only an amorphous structure is seen [Fig. 4(b)].

Recently many investigators have sought to isolate and characterize the so-called "structural" proteins from membranes. In this laboratory, a protein complex has been isolated from the synaptic and other membranes of brain tissue which has a strong affinity for ATP and other nucleoside triphosphates (7). The presence of this complex in synaptic membranes may account for the relatively high concentration of ATP associated with nerve endings. The binding of ATP to the protein does not require a divalent cation but appears to involve electrostatic interaction with basic

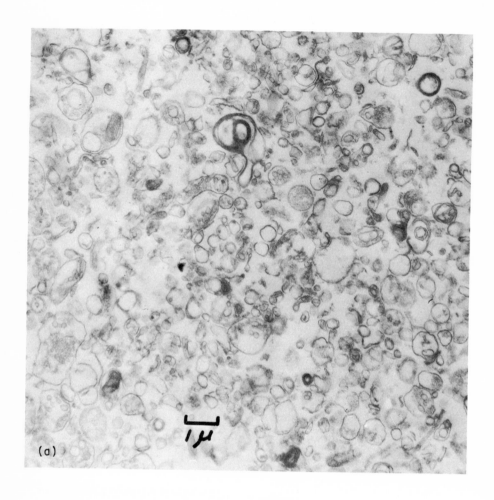

(a)

Fig. 2.   Electron micrograph of "microsome" fraction of rat brain.
(a) "Microsome" fraction prepared by centrifugation of a rat brain homo-
genate in 0.3 M sucrose at 50,000 x g for 30 min (after removal of larger
particulates at 10,000 x g).   Visible are fragments of endoplasmic reticulum, ribosomes, synaptic vesicles, nerve endings, neuronal-glial processes, myelin, and mitochondria; (b) Sonicated rat brain microsomal fraction
[See (7) for experimental details].

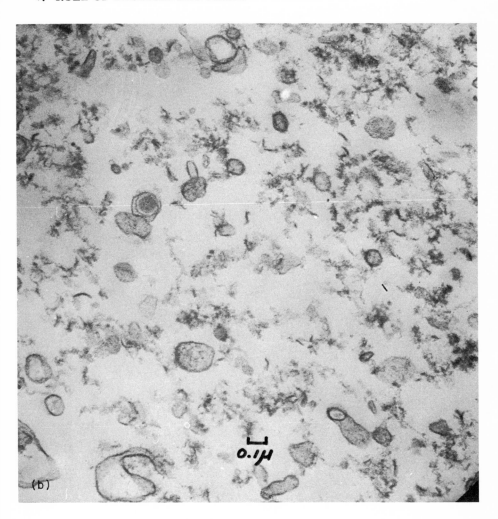

(b)

Fig. 2 (Cont)

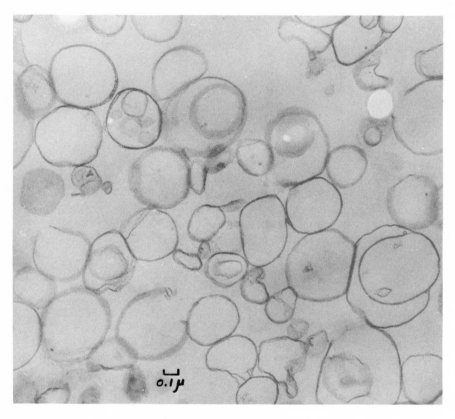

Fig. 3. Electron micrograph of isolated synaptic membranes. Membranes were prepared by subjecting isolated nerve endings to hypotonic shock and then separating the preparation to sucrose density gradient centrifugation. Magnification 50,000 x.

amino acids as well as van der Waal and, possibly, ion-dipole interactions. Although originally it was thought that the protein complex was not an ATPase, it now appears that it may exhibit such enzymic activity. Some of the other properties of the protein complex have been described elsewhere (7) and are referred to later. It is possible to form membrane-like structures with the protein mixture which have been examined ultrastructurally and chemically and utilized as a membrane model for biophysical studies.

From the electron micrographs the structural proteins appear to be arranged, for the most part, in the form of beaded chains of globular proteins in parallel arrays (Figs. 5,6). The appearance of the membranes formed from the purified proteins resemble closely those of isolated synaptic membranes (Fig. 5). At very high resolution the protein molecules appear to arrange themselves in the form of tubes having an inner diameter of 50-75 Å and an outer diameter of about 150 Å. If it be assumed that the estimate of the diameters of the globular proteins, namely 25-75 Å, is reasonable, then the proteins would have molecular weights ranging from $5 \times 10^4 - 20 \times 10^4$.

In their native state the structural proteins appear to exist in the form of aggregates of globular proteins arranged in parallel arrays in the form of a filament. X-ray crystallographic studies of proteins reveal that the globular configuration results from intramolecular interactions involving ionic, disulfide, and hydrogen bonds in addition to the weaker hydrophobic (van der Waals) forces. As is discussed later, when the protein is applied to an aqueous surface, it tends to uncoil (denature), arranging itself so that it assumes the shape of a single, extended polypeptide chain and has a limiting area of about 1 m^2/mg regardless of the protein (10, 11). In the compressed state the extended peptides form a solid, fibrous network which is rigid and completely water insoluble. In contrast to the globular proteins where intramolecular forces prevail, in the fibrous network (denatured state) intermolecular forces are preponderant.

In a classical monograph, Frey-Wyssling (12), in attempting to analyze the detailed fine structure of protoplasmic macromolecules, discussed the tendency for globular proteins to arrange themselves in beaded chains. The interaction between the globular proteins is presumably due to the integrating effect on two adjacent macromolecules of the repulsive electric double layer and attractive van der Waals forces. Such interactions are known to occur in highly concentrated gels of macromolecules (13). Frey-Wyssling (12) suggested that there existed a variable and discrete number of attractive loci in globular macromolecules accounting for the aggregation. Where only two such loci exist a beaded chain occurs, while the presence of three loci leads to the formation of a complex network with pores, four to the formation of tetrahedral arrays, and twelve to a close-packed crystalline lattice. To form a beaded chain or filamentous structure of the kind seen in the electron micrograph, it must be assumed that only a few attractive loci exist in the globular proteins.

If proteins were to comprise the basic matrix of at least some biological membranes, the question arises as to the location, molecular configuration, and relationships of the membranous lipids. As mentioned earlier, the classical view has assumed the lipids to be arranged in a bimolecular leaflet sandwiched between protein layers arranged in an extended fashion. An alternate view has been suggested in which the lipids are believed to project into the lipophilic core of globular proteins arranged in a beaded (2). On the basis of ultrastructural studies with biological and artificial

Fig. 4. Electron micrograph of pure lipid membranes. (a) Membrane prepared from a mixture of total lipids extracted from gray matter of rat brain. Membranes were prepared by applying lipid solution (in chloroform-methanol) to a solution containing Ca^{2+} and ATP [See (9) for experimental details]. Magnification 45,000 x. (b) Same as (a) except no Ca^{2+} was present. Magnification 45,000 x.

membranes, another variant of the Danielli-Davson model is possible. The inner membrane matrix may still be lipoidal, but the outer layers would be comprised of beaded chains of more or less globular proteins (Fig. 7). Such a model would have the dimensions of a biological (double) membrane and account for the structural features noted at high resolution. In the case of the artificial protein membranes, which appear to be comprised

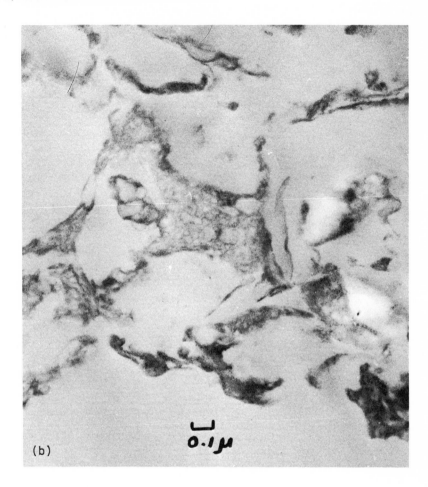

(b)

Fig. 4 (Cont.)

largely of tubular structures (Fig. 5), the inner core of the tubular components might be expected to contain lipids if, as suspected, the core is
lipophilic.

V. SOME PHYSICAL AND CHEMICAL PROPERTIES OF THE
STRUCTURAL PROTEINS

Since in the protein-lipid complex comprising the biological membrane
there are a number of different proteins varying in size, shape, and molecular weight, the overall structural configuration must be considerably
more complex than implied by the foregoing discussion. Although there

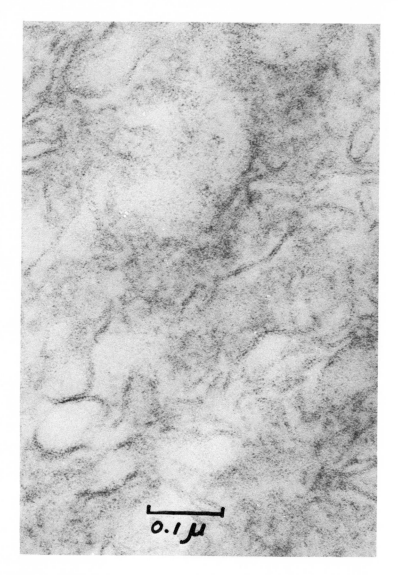

Fig. 5. Electron micrographs of artificial membranes prepared from structural protein. The primary structural unit appears to be tubular with the globular components (proteins) arranged in a parallel fashion. Occasionally a laminated arrangement is seen (L). See text and figure for explanation.

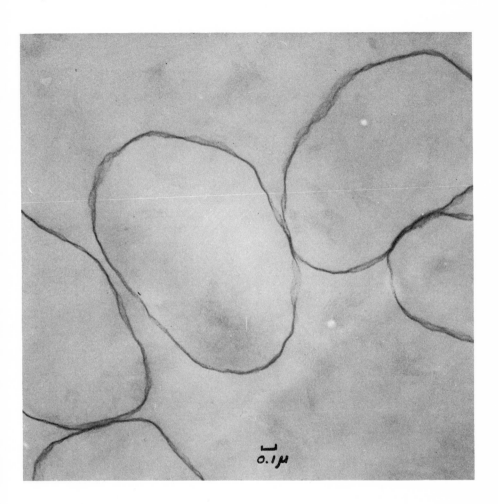

Fig. 6. Electron micrograph of artificial membranes prepared from solubilized lipid-protein mixture from rat brain synaptic membrane. After the synaptic membrane pellet (7) was homogenized and extracted twice in 0.05 M tris buffer, ph 7.5, it was extracted with 0.05 M potassium triphosphate (pH 11). The potassium triphosphate extract was dried as a thin layer on a petri dish at 25° for 2 hr. After the membrane was washed thoroughly in water, it was fixed in glutaraldehyde, dehydrated in ethanol, exposed to osmium tetraoxide, dried, and embedded in Epon, and sectioned. [See (6, 7) for details].

(a)   (b)

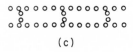

(c)

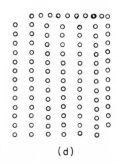

(d)

Fig. 7. Possible configuration of artificial membranes prepared from structural proteins. Interpretations based on electron micrographs (see Fig. 5): (a) Tubular arrangement of globular proteins in parallel array; (b) similar to (a) except that adjacent parallel arrays are occasionally linked by globular proteins; (c) ribbon-like arrangement comprised of two layers of proteins in a more or less parallel array. Although shown as being uniform, diameter of globular proteins vary from 25-75 Å

are regions in the electron micrographs of the structural protein complex where some uniformity is evident, other regions appear more heterogeneous in composition and may contain some uncoiled or extended proteins (Fig. 5). It is highly probable that the membranes formed from the purified proteins are comprised to some extent of denatured proteins irreversibly uncoiled to varying degrees. Nevertheless, the overall fine structure of the artificially prepared membranes is not very different from that of the isolated synaptic membranes (Fig. 3).

One of the unique and significant characteristics of the membranous proteins is their ability to readily interact to form high molecular weight complexes. The interaction is immediately evident once the lipids have been removed from the membranous complex. Such delipidized material is extremely insoluble, and even after solubilization the molecular weight

of the complex exceeds 10^6. If, however, the delipidized protein complex
is solubilized in 0.03 M sodium dodecyl sulfate (SDS) (pH9) and eluted from
a Sephadex gel filtration column with SDS, the highest molecular weight
detectable is 0.5 - 1 x 10^6 (14). After cleavage of protein disulfide linkages
by oxidation in performic acid (15), the molecular weight of the SDS-
solubilized material diminishes to the 10-50,000 range. When the SDS is
removed from the proteins (by ethanol or acetone), they once more tend to
aggregate to form a 10^6 molecular weight complex. After oxidation of the
disulfide groups, reaggregation, which is greatly reduced, can still occur
as a result of the combined action of hydrophobic forces and the diffuse
electric double layer of the proteins. As is discussed later, this capacity
of the proteins to form complexes or aggregates accounts for their structural role in the membrane.

A crucial problem concerning the properties of purified proteins is
the extent to which the purification procedures may have influenced the
native state of the protein. In their native state in neutral aqueous solutions
at lower temperature globular proteins exist in a state of minimal free
energy (16). When proteins in their original organized, helical configuration are altered by breaking H-bonds of S-S bonds, they assume a more dis-
organized random coil state. In many instances this helix-coil transforma-
tion is reversible, so that enzymic and other properties of the native protein can also be restored (17). Whenever a protein can readily undergo
helix-coil transformations it may be assumed that the primary structure
(amino acid sequence), rather than secondary and tertiary, alone determines both structural and functional characteristics of the protein.

The structural proteins derived from membranes of brain tissue appear
to retain their ability to bind ATP even after drastic procedures (e.g.,
boiling, strong acid-base treatment, urea treatment). Such procedures
would be expected to affect everything but the primary structure of
proteins. On the other hand, ATPase activity, particularly the Na-K activated component, is irreversibly affected after considerably milder treatment (e.g., few hours at room temperature). If it is assumed that the
protein exhibiting ATP-binding activity is also an ATPase, it must be concluded that the configurational specificity for enzymic activity is not required for ATP-binding. Furthermore, it appears likely that ATP-binding
involves intermolecular interactions and more than one type of protein.
Many proteins, particularly those not exhibiting enzymic activity, possess
"configurational adaptability," so that binding ability persists over a wide
variation in configuration (18). The ATP-binding proteins must undoubtedly
possess such adaptability.

It would appear from the foregoing discussion on the ultrastructural
features of membranes that proteins comprise an important part of the
membrane matrix. In view of the fact that a variety of enzyme systems

have been localized in membranes it is likely that such enzymes are an integral part of the matrix. Recent studies in this laboratory have demonstrated that the so-called "structural" protein of the synaptic membrane is comprised of a variety of proteins which exist in the form of aggregates held together by a variety of chemical and physical bonds. Although the exact number of distinct proteins involved is indefinite, electrophoretic studies reveal at least a dozen species (Fig. 8, B_1). Presented for comparative purposes are the electrophoretic patterns of the intact nerve membranes and membranes extracted from other subcellular components. It is readily seen that the patterns of the myelin (A), nerve ending membranes (B_1), and microsomes (M) are quite similar. As expected, more bands appear in the mitochondrial fractions (C,D) where there are numerous enzymes. Although the purified structural protein (not shown) appears to consist of only the electrophoretic band at the origin (visible in all patterns), there are actually a number of proteins represented. Their inter-

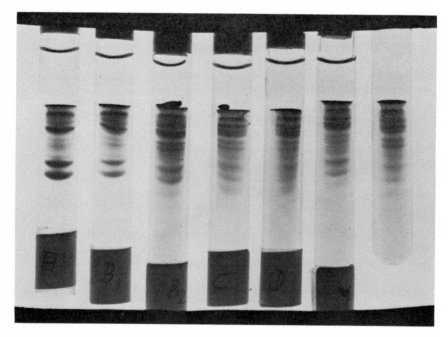

Fig. 8. Acrylamide gel electrophoretic patterns of structural proteins of various subcellular fractions of rat brain. My = myelin; B_1 = mostly membranes of nerve endings + myelin; B_2 = membranes of nerve endings; C = mitochondria, M = microsomes. All subcellular fractions were subjected to hypotonic shock, centrifuged at 50,000 x g, washed once with water, and recentrifuged. The residue was then dissolved in phenol: acetic acid: urea and run on disc electrophoresis.

action, which may involve hydrophobic, lyophobic, and disulfide bonds, is strong enough even in the presence of a hydrophobic solvent such as phenol so that the proteins cannot migrate in acrylamide gel. From amino acid studies of a number of the components it can be inferred that many of the proteins are comprised of a repeating subunit. When the protein complex is subjected to brief incubation with trypsin, at least 10 distinct peptides can be separated by paper chromatography, and the amino acid composition of many of the components is remarkably similar (Table 1).

TABLE 1

Amino Analysis of ATP-Adsorbing Protein and
Peptides After Trypsin Hydrolysis

	Original protein	Peptides (molar %)					
	Molar %	1	2	3	4	5	6
Asp	9.92 ±0.22	11.3	14.5	18.6	4.0	3.7	2.1
Thr	5.54 ±0.20	3.9	8.6	3.6	3.6	3.9	1.3
Ser	6.58 ±0.64	3.4	8.8	5.4	5.8	4.6	3.0
Glu	12.77 ±0.40	10.8	13.6	18.0	4.0	5.4	2.8
Prol	5.60 ±0.26	8.1	7.7	4.8	--	6.6	1.7
Gly	7.48 ±0.23	13.9	11.8	11.4	5.8	11.4	9.2
Ala	8.31 ±0.20	10.8	8.0	8.4	6.4	9.6	3.5
1/2 Cys	0.83 ±0.33	--	--	--	--	--	--
Val	6.06 ±0.11	5.8	3.5	3.6	12.0	7.2	1.6
Met	2.14 ±0.11	0.5	0.9	--	6.4	--	--
Ile	4.33 ±0.52	4.2	2.9	2.4	9.6	4.1	1.3
Leu	8.57 ±0.22	5.2	2.9	3.0	23.0	3.2	2.1
Tyr	3.14 ±0.13	1.6	1.2	0.6	3.8	1.0	--
Phe	3.85 ±0.15	2.1	1.8	0.6	6.6	1.0	1.1
Lys	7.55 ±0.40	11.0	7.7	16.2	4.4	21.7	44.5
His	2.26 ±0.10	3.1	3.2	2.4	2.0	1.6	--
Arg	5.05 ±0.22	4.2	2.9	1.2	2.6	16.9	25.4

TABLE 1 (Cont)

	Original protein	Peptides (molar %)					
	Molar %	1	2	3	4	5	6
Asp + Glu		22.1	28.1	36.6	8.0	9.1	4.9
Lys + His + Arg		18.3	13.8	19.8	9.0	40.2	69.9
Acid - basic		3.8	14.3	16.8	-1.0	-31.1	-65.0
Acid + basic		40.4	41.9	56.4	17.0	49.3	74.8
% Total protein	100	28	24	12	4	9	8

VI. BIOPHYSICAL MEASUREMENTS WITH PROTEIN MEMBRANES

One of the controversial questions of membrane physiology is whether the proteins or lipids are primarily responsible for the functional as well as ultrastructural characteristics of the membranes. Although both substances are major constituents of the membrane, their relative proportions vary greatly depending upon the membrane source. Furthermore, most membranes contain significant amounts of polysaccharides, particularly plant forms. Because of the heterogeneous composition and complex structural nature of membranes, interpretation of membrane structure from X-ray analysis, electron microscopy, and other analytical tools is difficult and laden with gross speculation. Long before such analytical tools were available, however, membranes, particularly from animal tissues, were known to be highly lipophilic, impermeable to many polar substances, and to have a high electrical resistance. Since such properties are characteristic of lipids, it has been inferred that the lipid constituents in the membrane are largely responsible for the biophysical properties of membranes, particularly in the case of the excitable membrane.

In recent years considerable interest has developed in the use of bimolecular lipid membranes as models for certain biological membranes, particularly those of neural tissue (19-21). The electrical resistance of such membranes has been reported to range from 10^6 -10^9 Ω-cm^2 and the capacitance from 0.4-1.3 μF cm^{-2} (19-21). The resistance of most biological membranes ranges from 700 Ω-cm^2 for squid axon to 4000 Ω-cm^2 for frog muscle; although somewhat lower than for pure lipid membranes, the resistance of biological membranes is still high compared to a dilute electrolyte solution. If a biological membrane is represented as a mixed electrical circuit consisting of resistances and capacitances, the capacity (plane plate condenser) can be described by the formula $C = D/4\pi d$, where D is the dielectric constant and d the distance between the plates of the condenser. If d is expressed in angströms and C in microfarads per square centimeter, the expression reduces to $C = 8.8 \ D/d$. By substituting values for C [obtained by high-frequency measurements (see Ref. 22 for discussion)] and assuming D to be about 10 within the membrane, the thickness d is calculated to be about 100 Å, which compares favorably with that for a bimolecular lipid membrane. The specific resistance (resistivity) of biological membranes would then be $10^9 - 10^{10}$ Ω-cm^2.

With the use of the structural proteins obtained from synaptic membranes it is possible to prepare membranes about 0.1-0.3 mm in diameter. Although such membranes are not stable enough for making electrical measurements, they are stable after drying and being coated with a thin layer of brain lipids. Membranes prepared in this manner had electrical resistances averaging 2000 Ω-cm^2, and the values remained constant regardless of the thickness of the lipid layered onto the protein membrane. The percentage lipid in the membrane (determined by chemical analysis) varied from 5-25%. Although it cannot be determined for certain what the contribution of the protein was to the resistance measurement, the amount of lipid present seemed to make little difference in the values obtained. The problem of whether the protein or lipid is responsible for the resistance measured is not as critical as that concerning the interpretation and significance of the resistance measurements.

Since the resistance does not vary with thickness of the membrane, it would appear that the values were largely due to the interfacial resistance. The electrical capacity of the membranes used in this study is similar to that of the double layer (solid-water interface), where the capacity is the result of the capacitance effects of the Stern and diffuse layers (23) acting in series and given by

$$\frac{1}{C} = \frac{1}{C_{\text{Stern}}} + \frac{1}{C_{\text{diff}}}$$

The capacity of the Stern layer (i.e., first plane of counter ions at a charged interface) is given by the formula for a condenser

$$C_{\text{Stern}} = 9 \ D/\delta \, \mu F \ cm^{-2}$$

where δ is the radius of the counter ion in angströms and D is the dielectric constant of the layer of counter ions. If the dielectric constant near the interface approaches 80, the value of water — a highly unlikely possibility —

so that the alkali cations are completely hydrated (i.e., $\delta = 4$ Å), C would have values ranging from 90 to 180 μF cm^{-2}. It is more likely that D is much lower at a highly charged interface, possibly around 10; therefore, actual values for C would range between 10-20 μF/cm^{-2}. At ionic concentrations in excess of 0.1 N, C_{diff} exceeds 70 μF cm^{-2} and would not (according to the equation) appreciably affect C. This value for the interfacial capacitance is an order of magnitude greater than that obtained for the capacitance of biological membranes (1 μF cm^{-2}); however, the value, which is based on certain tenuous assumptions, is an adequate approximation.

If the capacitance (and, accordingly, resistance) is due to the interface of membranes, it is understandable why the capacitance, within reasonable limits, does not appear to vary with membrane thickness. Furthermore, in biological membranes or artificial membranes constituted of biological materials, the capacitance would be reasonably constant, since the interfacial properties of lipids and proteins (e.g., potential, viscosity, and pressure) are in many respects similar.

Although electrical measurements of biological membranes, particularly excitable ones, are important parameters of biological function, a more direct index of membrane function is the flux of ions and other chemical substances across the membrane. Presumably, measurements of electrical resistance of impedance are a reflection of ionic fluxes, but the techniques of measurement (e.g., high frequency oscillations) may alter membrane structure. Since there was some question as to the compactness of the artificial membranes prepared from structural proteins, it was desirable to measure the extent to which radioactive substances could permeate. It was found that the membrane was largely impermeable to orthophosphate -^{32}P, ^{45}Ca, and labeled amino acids and sugars, while tritiated water penetrated slowly. It can be concluded that the membrane was not significantly more porous than an intact biological membrane.

VII. MEASUREMENTS ON SURFACE FILMS OF THE STRUCTURAL PROTEIN

A variety of physical measurements can be carried out with lipid or protein surface films which can reveal certain structural, electric charge, and other characteristics of the surface molecules. Included in these measurements are surface pressure, potential, viscosity, and adsorption, all of which have been helpful in investigating the interaction of Ca^{2+}, ATP, and various drugs with lipid and protein films.

Surface viscosity was measured by a ring torsion pendulum similar to one described by Kimizuka (24). A ring, 1.0 cm in diameter, was suspended from a copper alloy wire which had a torsion constant of 0.135 dyn cm. A mirror attached to the wire reflected a light onto a centimeter scale. The ring was lowered onto the surface of the subsolution, and a value for the surface viscosity was obtained by measuring the amplitude and the period of oscillation after a standard deflection of the wire.

Protein monolayers were spread from propanol-water (1:1) mixed solvent containing 0.5 M sodium acetate. Surface viscosity-area diagram for DEAE protein is shown in Fig. 9. According to Cumper and Alexander (25), who found that protein films begin to show the surface viscosity at

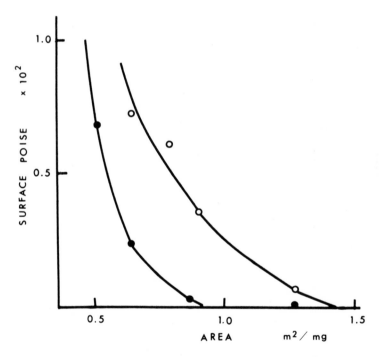

Fig. 9. Surface viscosity-area diagram of the structural protein. The protein was purified from rat brain synaptic membranes. To obtain differences with ATP, it was necessary to remove the cations bound to the protein by passing the protein solution through a Dowex-1 (anion-exchange) column. (●) = no ATP; (0) with 10^{-4} M ATP. At the limiting area (0.5-0.7 m^2/mg) the surface viscosity rises so rapidly that it cannot be measured. pH = 3.3, adjusted with HC1.

the area of close-packing, this protein has the limiting area of about 1.0 m^2/mg which is usually observed for proteins, irrespective of the kind of proteins (26).

The pH dependence on the surface viscosity of the protein was measured at an area of 0.509 m^2/mg (Fig. 10). The surface viscosity-pH curve is similar to that obtained for wheat gluten by Tschoegl and Alexander (27), although their measurement was carried out over a much higher viscosity

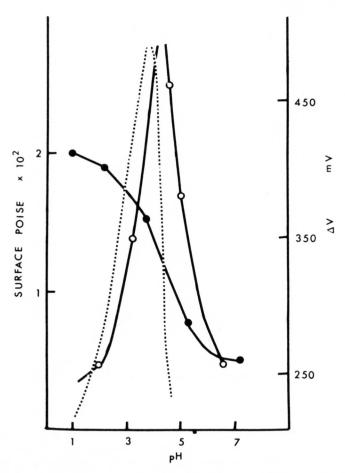

Fig. 10. pH dependence of surface potential, viscosity, and adsorption of structural protein film. (0) = surface viscosity, (●) = surface potential (mV), (----) = surface adsorption of ATP-C^{14} (10^{-4} M). Surface area was 0.51 m^2/mg.

region. The peak appeared around pH = 7.5 in their diagram and it was
regarded as the surface isoelectric point of wheat gluten where the maximum salt link is formed (27). Consequently, the protein should contain
more acidic amino acids (e.g., glutamic or aspartic acids) than basic
ones, and amino acid analysis subsequently confirmed this. The surface
potential measurement also showed a very striking change around pH = 4.2.
Since there is no basic amino acid which has the pK value around pH = 4.2,
this change must be attributed to the dissociation of carboxyl groups of
glutamic or aspartic acids which have pK's of 4.3 and 3.7, respectively.

From the surface viscosity-area diagrams of the ATP-protein it can
be seen that the viscosity of the protein is appreciably greater in the presence than in the absence of ATP (Fig. 9). One explanation for this observation is that the ATP, by interlinking basic sites on the proteins, may increase the size of the protein complex or alter its configuration. This
conclusion is also borne out by surface potential studies which show that
ATP greatly increases the surface potential of the protein complex. (A
greater compactness of the complex would increase the surface charge
density of the monolayer.) Other observations also support this conclusion, e.g., the greater turbidity of the protein solution in the presence of
ATP and the increased limiting area (force-area studies) of the protein
in the presence of ATP (7).

Since the pH optimum for ATP adsorption approximates that for surface viscosity, adsorption must be maximal near the isoelectric point of the
protein. Evidently basic amino acid residues are involved in the ATP adsorption, so that an excess of dissociated carboxyl acid groups can interfere. It is also possible that the shape and size of the protein complex is
so altered at the isoelectric point that ATP adsorption is also sterically
facilitated. It has been established that the nucleoside portion of ATP is
also involved in the adsorption (although neither AMP nor adenosine adsorbs
readily to the protein). In view of the fact that the shape and size of the
protein complex can influence surface viscosity, the pH-surface viscosity
curve may reflect more than the total ionic charge of the protein. Little is
known about the configuration of protein surface films, except that the
similarity in the limiting area of a variety of proteins implies that they
behave at the surface as single polypeptide chains (28).

The ionic nature of the ATP-binding to the structural protein is reflected in the effect of electrolyte concentration on surface viscosity and
ATP-adsorption (Fig. 11). Whereas the surface viscosity in the absence of
ATP increases with increasing concentration of NaCl, there is a decrease
in the presence of ATP. An increase in electrolyte concentration increases
the viscosity of a protein by a salting-out effect (i.e., dehydration + reduction in the zeta potential). Evidently ATP is able to prevent this effect

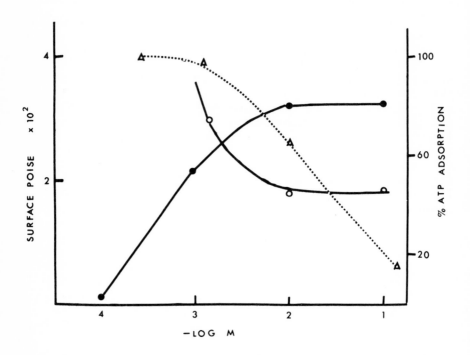

Fig. 11. Effect of electrolyte concentration on the surface viscosity and ATP adsorption of the structural protein. The protein was first passed through a Dowex-1 column. Surface viscosity without (●) and with (0) 10^{-4} M ATP. (Δ) = Surface adsorption of ATP-C^{14}. pH = 4.6. Viscosity was measured at a surface area of 0.64 m^2/mg. Adsorption of ATP-C^{14} was measured as described elsewhere (7).

because of the change in molecular configuration and electric charge distribution resulting from the adsorption of ATP to the protein. It is not clear why the surface viscosity in the presence of ATP does not return to a higher value at high electrolyte concentration, where the ATP adsorption is very low. Since the measurements on surface viscosity were conducted on protein monolayers while ATP adsorption was with a multilayered film, a strict comparison of adsorption and viscosity is not possible. The curve for surface viscosity versus electrolyte concentration in the presence of ATP does, however, bear resemblance to the surface potential-ionic strength curve which is measurable with monolayers.

From a study of the surface viscosity of the structural proteins it can be seen that the pH-dependent curve for ATP adsorption corresponds with that for surface viscosity (Fig. 10). Since the surface viscosity of a protein is maximal at the isoelectric point, where intermolecular and intramolecular interactions are optimal, it might be inferred that the surface viscosity-pH curve is a reflection of the net electrostatic charge on the protein complex. The surface potential-pH curve (Fig. 10), indicates that the pK of the surface film is approximately 4.5 which is in the range of pK values for glutamic and aspartic acid residues of proteins. From the surface potential-pH curve it can be seen that the surface COOH groups are not fully dissociated until the pH exceeds 7. It is likely, however, that factors other than the ionic charge were influencing the surface viscosity of the protein complex. Included in these would be the size and shape of the protein complex at the surface. Little is known about the configuration of protein surface films, except that the limiting area of the structural proteins is the same as that of many other proteins, such as globulins, albumins, and hemoglobin (25, 27).

A comparison of the ATP-binding activity of various protein fractions derived from beef brain indicates that the more insoluble, high molecular weight complexes exhibit the greatest activity. This observation would suggest that ATP adsorption is optimal when the protein complex is aggregated.

Although ATP adsorption is not seen when the pH of the bulk phase is above 5, the pH at the surface could be higher by as much as two pH units. The surface pH is related to the bulk pH by the following expression (23, 29):

$$pH_s = pH_b + \epsilon \psi/kT$$

when pH_s is the surface pH, pH_b is the bulk pH, ϵ is the electronic charge, ψ is the potential of the diffuse ionic double layer (equivalent to surface potential), k is the Boltzmann constant, T is the absolute temperature. From this expression, when the surface potential is +200 mV, pH_s would be two units greater than pH_b (pH_s would be correspondingly less than pH_b if ψ was negative). It is conceivable, therefore, that ATP adsorption occurs when the pH at the interface is near 7, as would be the case in a biological membrane. Another point of interest is the fact that the surface adsorption of ATP accounts for an additional 200-300 mV increase in the surface potential, so that pH_s may be still higher. An increase in ψ with adsorption occurred despite the fact that ATP is strongly anionic.

The surface potential, ΔV, is a function of the number of molecules, n, per unit area of surface, with a net dipole, μ, plus the electrostatic charge, ψ_o, in the ionic diffuse double layer, given by the expression

$$\Delta V = 4\pi n \mu + \psi_o$$

In a surface film, ΔV is a measure of the sum total of dipoles and electro-static charges. To explain the net increase in the ΔV resulting from ATP, it must be assumed that the anionic charge of the phosphate groups is either offset by the dipoles of the adenosine and phosphate groups or that the configuration of the phosphate groups within the double layer is such as to contribute a negligible anionic charge. It is also possible that the inter-action of ATP with the protein results in a condensation of the film, thus increasing the charge surface concentration, n. ATP could either affect the intramolecular or intermolecular configurations of the surface protein.

VIII. LIPID-PROTEIN-ATP INTERACTIONS

A study of the cytoplasmic distribution of ATP in brain tissue revealed that a considerable portion of the ATP was associated with membranes, par-ticularly those derived from nerve endings (6). Initially it was thought that the ATP may have been bound to lipids in the form of an ATP-Ca-phospho-lipid complex (32). It was subsequently shown that the membranous ATP was associated largely with the protein. Extraction of the lipids with organic solvents did not remove any nucleotides but removed most of the membrane-bound Ca^{2+}. Although there is no doubt that proteins are the major site of ATP, it was of importance to determine the effect of lipids on ATP adsorption since membranous protein must exist as a complex with lipids.

There are a variety of ways of testing the effect of lipids on ATP ad-sorption. One way is to determine the effect of lipid removal on the ATP adsorption to sonicated membrane preparations; another is to work with emulsifications of the purified membrane proteins with various lipids. Still another method involves the use of mixed surface films of lipids and proteins. By removal of the lipids from the sonicated membranes with organic solvents, over half of the total ATP-adsorptive activity remains (7). On the other hand, if the extracted lipid is emulsified (by sonication) with the remaining sonicated material, so that the original proportions are obtained, most of the activity is restored.

The studies with mixed films of the purified protein and various lipids have not proved very useful, since interactions of the lipids and proteins at the surface do not occur readily. The presence of one substance at the interface tends to exclude the other, whether protein or lipid is used first. It is, however, possible to measure the interaction of the protein with ATP in the bulk phase utilizing this technique.

With a basic lipid such as octadecylamine surface adsorption of ATP readily occurs when the pH is on the basic side. In the presence of a highly basic protein, such as protamine, adsorption to octadecylamine is prevented, while with a protein such as gamma globulin, whose pK is acidic, there was no effect. The adsorption of ATP to a mixed surface film of the structural protein and octadecylamine is greater than with octadecylamine alone. Other long chain amines (e.g., cetyltrimethylammonium chloride or cetylpyridinium chloride) have an action similar to that of octadecylamine.

IX. THE ROLE OF MEMBRANOUS ATPases

The suggestion that ATPases may be involved in bioelectric phenomena had been proposed a number of years ago (30), but the manner in which it participates is still largely obscure. In more recent years with the discovery of Na^+-K^+ dependency of certain membrane-bound ATPases the concept has arisen that the ATPases are iogenic systems for transferring Na^+ and K^+ across cellular membranes [see Tanaka (31) for review]. Although it is likely that a structural or conformational change in the enzyme system is involved in determining ion selectivity and transfer, little light has been shed on this problem.

An important fact concerning its role is the phospholipid-dependency of the system, suggesting that the enzyme system exists as a lipoprotein or protein-lipid micellar complex within the membrane. A simplified molecular model of such a complex has been proposed (9). In this model Ca^{2+} forms the link between the ATP and phospholipid, while the enzymic properties of the enzyme are associated with Mg^{2+} (Fig. 12). Since K^+

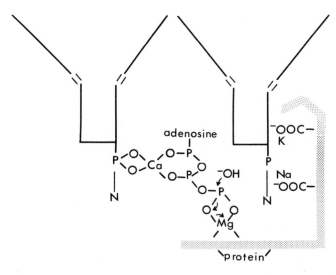

Fig. 12. Molecular model for an excitable membrane complex involving Ca^{2+}, ATP, phospholipd-cholesterol, and the Na^+, K^+, Mg^{2+}-ATPase system. The alkali cationic sites may be COO^- groups of acidic amino acids.

is less hydrated than Na^+, it is presumed that it is associated with the lipophilic position of the protein (or proteins) which project into the lipophilic region of the lipids. It has been conjectured that the dissociation of the Ca^{2+} from the complex is a factor determining in part the susceptibility of ATP to hydrolysis (9, 32).

In view of the more recent work with the structural proteins of the synaptic membrane -- a major site of the Na, K^+-ATPase -- it would appear that the proposed model is greatly oversimplified. As discussed earlier the structural proteins consist of a number of different proteins one or more of which exhibit ATPase activity. Although the protein site for ATP storage is distinct from that for enzymic activity, the latter being Mg^{2+} - dependent, it is to be expected that a close relationship exists between the two sites. The high concentration of ATP in synaptic and other neuronal membranes would require a separation of the two sites; however, the protein-lipid complex may undergo a rearrangement permitting their association. As discussed in Tanaka's contribution to this volume (31), the so-called Na^+, K^+, Mg^{2+}-ATPase is an enzyme complex and could be comprised of a number of proteins. It is possible, therefore, that this complex is the same structural component of the membrane which exhibits ATP adsorption.

X. A MOLECULAR MODEL FOR THE EXCITATORY MEMBRANE

A membrane model of an excitatory membrane has been proposed in which the functional unit is a complex of Ca-ATP-phospholipid-protein (9, 32). One molecular model has been proposed which assumes the Ca^{2+} to bind both to ATP and the phosphate group of a phospholipid (Fig. 12). Included in this model is a protein (or protein complex), which could conceivably be an ATPase (possibly an ion-specific ATPase, responsive to both K^+ and Na^+). The evidence in support of this model has been reviewed elsewhere (9) and only certain features of it will be described. As discussed earlier, the first chemical event associated with depolarization is the displacement of the Ca^{2+} from the complex. This is followed by the release of ATP and a structural change in the membrane which permits Na^+ influx. ATP then becomes vulnerable to hydrolysis, although the cleavage reaction (ATP + enzyme $\rightleftharpoons$ enzyme ~ P + ADP) may be reversible with a minimal loss in free energy (32). Repolarization occurs with the recombination of Ca^{2+} which, in turn, may be dependent upon the regeneration of ATP. Excitation is possible as long as ATP is available for recombination; and when the total cellular concentration of ATP decreases below 75%, excitation fails (32).

One of the basic problems of the model concerns the role of ATP. As discussed previously, ATP can modify the physical characteristics of structural proteins and, by virtue of its chelating and sequestering action, influence the availability of Ca^{2+}. Besides influencing the intramolecular configuration of structural proteins, ATP could affect their interaction with other proteins and lipids. It can, by regulating electrostatic and other charge characteristics of the membrane, influence the flux of ions and other substances across the membrane.

Elsewhere (32) the evidence has been reviewed for the importance of ATP and phosphorylation in bioelectric phenomena and the manner in which various metabolic inhibitors influence membrane function. More direct evidence for the role of ATP on membrane function has only recently become available. ATP and other nucleotides have been shown to influence the excitatory characteristics of nerve and muscle tissues (33). In frog nerves ATP antagonizes the spontaneous excitation produced by Ca^{2+} removal; while in the absence of added ATP, the concentration of Ca^{2+} needed to stabilize excitability is 10-100 times greater than in its presence (34). In smooth muscle ATP, AMP, and adenosine decreased the spontaneous tone, the frequency of spike discharge, and hyperpolarization, ATP being a 100-fold as potent an inhibitor of spontaneous activity as AMP. ATP and other adenosine derivatives also prevented the intradepolarization induced by acetylcholine, while relaxing the muscle after K^{+}-induced contracture (35). These observations are consistent with the hypothesis that the adenine nucleotides influence membrane properties. With the use of metabolic inhibitors it could be shown that the contractility of heart cells grown in tissue cultures is dependent upon the actual concentration of ATP; and that a decrease below 85% of the normal ATP concentration leads to an inability of the cells to contract in response to stimulation (36).

XI. AGENTS AFFECTING MEMBRANE Ca^{2+}

The process of excitation in biological systems is attributed to the transient flow of ions across a membrane which, in its resting state, is able to maintain an electrochemical ionic gradient. There is now general agreement that Ca^{2+} plays a key role in the structure and function of the excitable as well as other types of biological membranes. It has been proposed that membranous Ca^{2+} exists in the form of a complex with ATP, phospholipids, and proteins and that the dissociation of Ca^{2+} from the complex is the critical event leading to transient increase in Na^{+} flux responsible for the action potential [see (32) and (36) for review]. Among the important problems relating to the excitable membrane is the chemical nature of the Ca^{2+}-membrane complex and the mechanisms leading to the transient changes associated with ionic fluxes.

Since Ca^{2+} plays such a fundamental role in the regulation of membrane permeability, it is to be expected that agents which can replace, displace, or complete with Ca^{2+} in the membrane complex would affect membrane function. Examples of agents which can displace Ca^{2+} are chelating agents such as EDTA, citrate, and polyphosphates. Among the agents which are known to compete with Ca^{2+} are the local anesthetics and various anti-cholinergic drugs. Included in this latter group are a class of glycolic acid esters possessing potent psychotomimetic action. Since most neuro-tropic drugs act directly on the excitatory mechanisms of neurons, their locus of action would be on synaptic, axonal, or other excitatory mem-branes. The glycolate esters are localized primarily in the membranous components of brain tissue (37) and appear to be associated with lipids. With the use of lipid surface films and micelles it can be demonstrated that the esters have striking effects on such properties as surface viscosity, tension, and potential; and with tritiated esters, it could be shown that the agents not only strongly adsorb to lipids but are themselves surface active (37,38). Such studies reveal that the nature of the interaction between the drugs and lipids bears some resemblance to that occurring between Ca^{2+} and lipids. Furthermore, it could be shown that the glycolate esters in-teract with ATP alone or in combination with lipids, an observation which would suggest that the drugs could also interfere with the role of ATP in the membrane.

In some instances it is possible to directly correlate the effects of the drugs in surface films with those on the bioelectric properties of excitable tissues. If frog sartorius muscles are immersed in a Ca^{2+}-free Ringer's solution, the resting membrane potential of the individual muscle fibers decreases from -90 to -60 mV. Reimmersion of the muscles in a normal Ringer's solution containing 2 mM Ca^{2+} will restore the potential to its normal value. Similarly, reimmersion of the muscles in Ca^{2+}-free solu-tion containing 10^{-4} -10^{-5} M of a glycolate ester (e. g. , N-methyl 4-piperidyl benzilate) will also restore the resting potential. This experi-ment cannot, however, be repeated on frog spinal ganglion cells. It can be concluded, therefore, that in the case of the sartorius muscle the drugs are simulating the action of Ca^{2+}. At larger concentrations (10^{-3} M) or for prolonged exposure at lower concentrations, the glycolate esters block nerve conduction as do procaine and other local anesthetics (7). Other electrical effects of the glycolate esters are blockade of the positive after-potential and prolongation of the negative after-potential of frog spinal ganglion cells and sartorius muscle fibers (7). Such phenomena are due to the ability of the drugs to "stabilize" the excitable membrane, presumably because the drug-membrane complex is not as readily dissociable as the Ca^{2+}-membrane complex (38).

Some agents affect membrane function by increasing affinity for Ca^{2+}. The binding of ^{45}Ca by cardiac microsomes was increased by both ouabain (10^{-7} -10^{-4} M) and 4-hydroxy-3-pentenoic acid a -lactone (10^{-3} M) in the presence of 5 mM ATP but was unaffected in its absence (39). Both these agents augmented Ca^{2+} turnover and increased the activity of Ca^{2+}-stimulated ATPase. From these observations the authors concluded that the two unsaturated lactones with cardiotonic activity exerted their effects by increasing the availability of Ca^{2+} needed for contraction.

Another way to affect membrane Ca^{2+} is to alter the macromolecular components by enzymes. Phospholipase c (which cleaves the glyceride-phosphate bond of phospholipids) will produce rapid Ca^{2+} loss and irreversible depolarization (32). Trypsin, which modifies the lipoprotein structure of microsomal membranes of rat muscle, as evidenced by X-ray diffraction and electron microscopic examination, produces a loss of ability to accumulate Ca^{2+} (40). The primary site of Ca^{2+} storage is presumably in vesicular components of the sarcoplasmic reticulum.

XII. SOME OTHER AGENTS AFFECTING THE EXCITATORY MEMBRANE

A number of cyclopeptides and cycloethers have been shown to facilitate cationic transfer across biological and artificial lipid membranes. Such compounds, which are highly lipophilic, are capable of binding specific cations (e.g., K^+) by ion-induced dipole interactions, and by so doing permit their transport through the lipophilic region of the membrane interior. Evidently, the partial neutralization of the cation in the caged interior of the macrocyclic compound minimizes its interaction with water dipoles, making it more compatible with a region of low dielectric. Among the interesting properties of such macrocyclic molecules is their ionic specificity; e.g., K^+ is preferred over Na^+ in the case of valinomycin where the internal cubic array formed by eight oxygen atoms better accommodates K^+ than Na^+ (41). With the use of bimolecular lipid membranes a number of macrocyclic compounds can be shown to accelerate the transport of K^+ (42); however, such compounds have no significant effect on ion transport across protein-lipid membranes. One possible explanation for this difference in the behavior of the two types of membranes is their difference in electrical resistance or lipophilicity, as described earlier. It has been estimated (43) that the so-called bimolecular phospholipid membranes may be comprised of as much as 70% solvent (usually n-decane), consequently many of the properties of such membranes may be largely due to the solvent. Resistance measurements of "membranes" prepared from n-decane in a similar manner have values ranging from 10^7-10^9 Ω-cm^2 (44).

Various macrocyclic compounds exert a highly specific action on certain enzyme systems involved in protein, nucleic acid, and mitochondrial

ATP synthesis. It is likely that many of these substances act upon the
membranous organization of enzymes by affecting directly or indirectly
(by interacting with lipid co-factors) enzyme activity. Atractyloside exerts
its action on mitochondrial electron transport coupled to ATP synthesis
by preventing the binding of ATP (or ADP) (45). Evidently the interaction
of atractyloside with a membranous component of mitochondria -- at a site
which is different from that responsible for ATP-binding -- results in a
conformational change preventing either the binding or, after binding, the
loss of ATP. To what extent the macrocyclic compounds are affecting
mitochondrial ATP synthesis by virtue of their ionophoric action is not
known. It is plausible that if an enzyme does have a requirement for an
inorganic cation, particularly at a lipophilic site, the presence of a
macrocyclic ionophore can then influence enzyme action.

There is an interesting class of compounds having a highly specific
action on the membrane conductance of the axonal membrane (46) and they
appear to block the peak transient component (i.e., action potential) of
membrane conductance but not the steady-state (resting) component. Included among these agents is tetrodotoxin which is a poison of the puffer
fish. In view of the fact that an agent like DDT -- which also blocks
nerve conduction -- delayed the off process of the peak transient of Na
conductance, it was concluded that operationally separate ionic channels
exist in the axonal membrane. To date nothing is known about the molecular mechanisms accounting for the action of such compounds. It is obvious
that before mechanisms can be proposed there is a need for understanding
the molecular configuration of the excitatory membrane, particularly regarding the sites of Na^+ transport.

XIII. CONCLUSION

The present discussion has focused primarily on the complex, dynamic
nature of the membrane, particularly that derived from excitatory tissues.
To a considerable extent biogenic amines and most pharmacological agents
affecting the nervous system probably exert their primary action on the
excitatory membrane. It has been suggested that the structural proteins
of the membrane, by virtue of their capacity to reversibly interact with
one another and undergo conformational changes, may be a major one in
determining the structural and functional characteristics of the membrane.
One working model for the excitatory membrane has been proposed which
is based on the premise that Ca^{2+} and ATP are responsible for the reversible structural changes regulating ion flow accompanying excitation. A few
examples of agents directly influencing the interaction of Ca^{2+} and ATP
with the lipoprotein structural complex have been presented, and it is
conjectured that a wide variety of other agents, including biogenic amines

and other neurohumors, may be acting on this system. Although there are
numerous difficulties working with artificial membranes and surface films,
they nevertheless have sufficient similarities to natural membranes in that
they are extremely useful in the investigation of membrane phenomena.

REFERENCES

1. J. F. Danielli and H. Davson, J. Cellular Comp. Physiol., 5, 495
   (1935).
2. F. Sjostrand, Protoplasma, 63, 248 (1967).
3. F. Sjostrand, Abstracts 155th American Chemical Society Meeting,
   1968.
4. R. Tanaka and L. G. Abood, J. Neurochem., 10, 571 (1963).
5. V. P. Whittaker, Prog. Biophys. Mol. Biol., 15, 39 (1965).
6. L. G. Abood, K. Kurahasi, E. Gruner, and M. P. del Cerro,
   Biochim. Biophys. Acta, 153, 531 (1968).
7. L. G. Abood and A. Matsubara, Biochim. Biophys. Acta, 163, 539
   (1968).
8. W. D. Stein, The Movement of Molecules Across Cell Membranes,
   Academic, New York, 1967.
9. L. G. Abood, in Neurosciences Research (S. Ehrenpreis and O. C.
   Solnitzky, eds.), Academic, New York, 1969, p. 42.
10. A. E. Alexander and T. Teorell, Trans. Faraday Soc., 35, 727 (1939).
11. J. T. Davies, Trans. Faraday Soc., 49, 949 (1953).
12. A. Frey-Wyssling, Submicroscopic Morphology of Protoplasm,
    Elsevier, Amsterdam, 1953.
13. G. Z. Oster, Electrochem. U. Angew Physik. Chem., 55, 529 (1951).
14. M. Reiter, H. Sarau, and L. G. Abood (unpublished).
15. F. Sanger, Biochem. J., 44, 126 (1949).
16. J. T. Edsall, in Structural Chemistry and Molecular Biology (A. Rich
    and N. Davidson, eds.), Freeman, San Francisco, 1968, p. 88.
17. C. B. Anfinsen, Harvey Lectures, Ser. 61 (1965-1966), p. 95.
18. J. F. Foster, M. Sogami, H. A. Petersen, and W. J. Leonard,
    J. Biol. Chem., 240, 2495 (1965).
19. P. Mueller and D. O. Rudin, Biochem. Biophys. Res. Comm., 26,
    398 (1967).
20. D. A. Haydon, J. Am. Oil Chemists' Soc. 45, 230 (1968).
21. T. Hanai, D. A. Haydon, and J. L. Taylor, Proc. Roy. Soc. (London),
    A281, 377 (1964).
22. K. S. Cole, Membranes, Ions, and Impulses, Univ. California,
    Berkeley, 1968.
23. J. T. Davies and E. K. Rideal, Interfacial Phenomena, Academic,
    New York, 1961, p. 101.
24. H. Kimizuka, Bull. Chem. Soc. Japan, 29, 123 (1956).

25. C. W. N. Cumper and A. E. Alexander, Trans. Faraday Soc., 46, 235 (1950).

26. T. Isemura, Colloidal Surfactants, Academic, New York and London, 1963 (No. 2), p. 270.

27. N. W. Tschoegl and A. E. Alexander, J. Colloid Sci., 15, 168 (1960).

28. K. Shinoda, T. Nakagawa, B. Tamamuski, and T. Isemura, Colloidal Surfactants, Academic, New York, 1963.

29. J. J. Betts and B. A. Pethica, Trans. Faraday Soc., 52, 1581 (1956).

30. L. G. Abood and R. W. Gerard, J. Cellular Comp. Physiol., 43, 379 (1954).

31. R. Tanaka, this volume.

32. L. G. Abood, Intern Rev. Neurobiol., 9, 223 (1966).

33. J. Axelsson and B. Holmberg, Acta Physiol. Scand., 75, 149 (1969).

34. A. S. Kuperman, M. Okamoto, and E. Gallin, J. Cellular Comp. Physiol., 70, 257 (1967).

35. M. W. Seraydarian, I. Harary, and E. Sato, Biophys. Biochim. Acta, 162, 414 (1968).

36. K. Koketsu, in Neurosciences Research (S. Ehrenpreis and O. C. Solnitzky, eds.), Academic, New York, 1969, p. 1.

37. L. G. Abood and J. H. Biel, Intern Rev. Neurobiol., 4, 218 (1962).

38. L. G. Abood, in Drugs Affecting the Central Nervous System (A. Burger, ed.), Vol. 2 Dekker, New York, 1968, p. 169.

39. M. L. Entman, J. W. Cook, and R. Bressler, J. Clin. Invest., 48, 229 (1969).

40. R. Coleman, J. B. Finean, and J. E. Thompson, Biochim. Biophys. Acta, 173, 51 (1969).

41. B. C. Pressman, Federation Proc., 27, 1283 (1968).

42. G. Eisenman, S. M. Ciani, and G. Szabo, Federation Proc., 27, 1289 (1968).

43. F. A. Henn and T. E. Thompson, Ann. Rev. Biochem., 38, 241 (1969).

44. R. Waldbillig and A. Matsubara (unpublished).

45. H. H. Winkler and A. L. Lehninger, J. Biol. Chem., 243, 3000 (1968).

46. T. Narahashi and J. W. Moore, J. Gen. Physiol., 51, 93S (1968).

Chapter 2

BIOPHYSICAL ROLE OF Na$^+$, K$^+$, Mg^{2+} -ACTIVATED ADENOSINE TRIPHOSPHATASE IN NERVE CELL MEMBRANE

Ryo Tanaka

Center for Brain Research
University of Rochester
Rochester, New York

I. INTRODUCTION

Since a Na^+, K^+, Mg^{2+}-activated adenosine triphosphatase (Na, K, Mg-ATPase) system in the microsomal fraction of crab nerve fiber was first assumed by Skou (1) to be playing a central role in active ion transport across the cell membrane, considerable evidence for this hypothesis has accumulated. It is now well established that there is an ATPase activity associated with active transport, which accumulates K^+ and expels Na^+ to maintain the high K^+ concentration and the low Na^+ concentration in the cell. Since there are many detailed reviews on the subject from various points of view (2-7), the present discussion is limited mainly to the facts relevant to the activity of the cell membrane of the nervous system.

II. RELATION BETWEEN Na, K, Mg-ATPase AND ION TRANSPORT

A. Ion Transport System

K^+ ions present at a high concentration in the cell do not seem to be counter ions of the fixed, negatively charged groups of colloidal protein inside the cell. The physiological resting and action potentials can be recorded from squid axons which have been perfused with salt solution containing no negatively charged macromolecules to attach K^+ and lack original axoplasm (8, 9). The diffusion coefficient and mobility of K^+ in the axon in an electric field were almost the same as those in sea water [see (10)] and the activity coefficients of cations in axoplasm were not as low as would be expected from the interaction of the cations with fixed negative charges (11). Thus, it seems to be reasonable to assume an ion transport pump associated with the cell membrane separating the protoplasm from its milieu.

B. Similarities in Localization and Properties

The basis of the belief that an ion pump is closely associated with the Na, K, Mg-ATPase system may be that both systems share numerous characteristics.

1. Localization.

The strategically best localization of an ion pump is the envelope of the cell. This idea has been supported by the fact that reconstituted red cell ghost as well as giant squid axon internally perfused were capable of accumulating K^+ and extruding Na^+ if ATP or creatine phosphate was in the structures (8, 9). As is discussed in the later sections, there are many convincing evidences, supported by both biochemical and histochemical techniques, to believe that the ATPase system is present in the cell membrane. In the case of the nervous system, specifically high activity of the enzyme was found in the membrane of the neuronal element.

2. Dependence on Metabolism and ATP.

Inhibition of respiration of squid axon, brain slice, and retina by cooling, deprivation of oxygen, and administration of inhibitors of oxidative phosphorylation, induced a net gain of Na^+ and net loss of K^+ (12-16), and in some other tissues both oxidative phosphorylation and glycolysis had to be inhibited to result in the same ion distribution. These observations may lead one to assume that the pump is driven at the expense of energy derived from splitting a terminal phosphate from ATP, a high energy compound common to both metabolic processes.

Na^+ efflux of the squid axon whose oxidative phosphorylation was stopped by dinitrophenol fell in parallel relation with the decrease in ATP level in the axon, and the injection of ATP, arginine phosphate, or phosphoenolpyruvate into the axon partly restored Na^+ efflux (17, 18). The latter two high energy phosphate compounds were believed to work via ATP formation by transferring phosphate to ADP. This was confirmed by Brinley and Mullins (19), who tested nine naturally occurring high energy phosphate compounds to find out that only ATP could support 80-100% of normal Na^+ extrusion from internally dialyzed squid axon (20, 21), and that arginine phosphate could replace ATP only when ADP was present in the axom (20).

On the side of Na, K, Mg-ATPase, nucleotide specificity of a partially purified ATPase preparation of rabbit brain was reported to be specific only to ATP and CTP, whose hydrolysis rate was 15% of that of ATP, and no nucleotide diphosphate was hydrolyzed (22). Thus the fuel of the pump and the substrate of the ATPase are common.

3. Coupling of Na^+ and K^+.

Both ion pump and Na, K, Mg-ATPase system possess the spatial orientation required for active transport. It is known that Na^+ efflux from squid axons was reduced when external K^+ concentration was lowered (13, 20, 23, 24), and K^+ influx was doubled by a 5 to 6-fold increase in internal Na^+ concentration (25). The overall form of a curve relating Na^+ extrusion to external K^+ concentration was S-shaped (25). The dependence of sodium pump on external K^+ was indirectly proved with desheathed vagus nerves of rabbit (26, 27), and half maximal activation of the pump of the vagus was obtained by 3.8 mM of external K^+ and 29.5 mM of internal Na^+, which was replaceable by Li^+ although its effectiveness was low (28).

Using reconstituted ghosts of human red cell Na, K, Mg-ATPase was similarly shown to be activated only when Na^+ was present in the internal and K^+ in the external fluid (29). Orthophosphate production in an intact crab nerve was stimulated by internal K^+ and external Na^+ (30). The ATPase of membrane preparation requires high Na^+ concentration and low K^+ concentration for optimum activity, and Na^+ can be replaced by Li^+ as in the case of the pump of rabbit vagus nerve mentioned.

4.   Inhibitors.

Ouabian has long been known as a potent specific inhibitor of the ion transport of a variety of tissues including nerve tissue (21, 24, 26, 31), while the cardiac glucoside was also found to be a specific inhibitor of Na, K, Mg-ATPase of many tissues. The alteration of its inhibitory potency on the enzyme activity induced by molecular modification paralleled the change of its inhibitory efficacy on ion transport (32-34). Ouabain inhibited the transport and the enzyme acvitivy only when applied on the outer surface of the membrane (34-36), and the inhibition of both activities was reduced by augmentation of external K^+ concentration (37, 38).

Oligomycin showed effects similar to those of ouabain on the various aspects of active transport. It inhibited ion transport in rat liver slices, whether the transport was dependent on respiration or on glycolysis (39), and it inhibited the transport, Na, K, Mg-ATPase, and lactate formation controlled by the ATPase in red cell ghost (40, 41).

5. Others.

Garrahan (42) loaded red cell ghosts with a large quantity of K^+, ATP, ADP, [32]Pi, and a little Na^+, and the ratio (ATP)/(ADP)·(Pi) was kept low, using the reconstitution technique. When the loaded ghosts were incubated in high-sodium, potassium-free media there was an extra incorporation [32] Pi into ATP, and the extra incorporation was largely abolished by ouabain and partly by oligomycin. This interesting observation may be interpreted as showing that the transport system ran backyards at a measurable rate and ATP was synthesized at the expense of energy derived from ionic concentration gradients.

Studies on some pathological conditions suggest a close correlation between the two systems. Neonatal hypothyroidism increased Na^+ and Cl^- contents and decreased K^+ and Mg^{2+} in both cerebral cortex and cerebellum of developing rat, and the changes were accompanied by a diminishing of the activity of Na, K, Mg-ATPase of the brain (43). Adrenalectomy of rats resulted in decreased active ion transport of kidney and at the same time decrease in kidney Na, K, Mg-ATPase activity was observed. The electrolyte contents of brain were not affected by the operation, however, and no change in the enzyme activity of brain was found (<u>44</u>).

C. General Properties of ATPase

1. Intermediate.

When ATP-γ-^{32}P was incubated with Na, K, Mg-ATPase preparation from a number of tissues under different conditions, various amounts of incorporation of radioactivity into protein fraction were noted. The incorporation was high in the presence of Na^+ and Mg^{2+}, and was lowered by the further addition of K^+. The action of K^+ was reversed by a low concentration of cardiac glucosides or oligomycin (<u>45-50</u>). It appears as if the terminal phosphate of ATP were transferred to some acceptor(s) in the enzyme system to form an intermediate(s) and then the bound phosphate was released as Pi, this step being activated by K^+ and inhibited by the cardiac glucosides or oligomycin. The nature of the phosphorylated intermediate has been suggested to be an acyl phosphate (<u>47</u>, <u>48</u>, <u>51</u>).

2. Reaction scheme.

Based on his own experiments and others, Albers proposed the following scheme (<u>5</u>):

$$E_1 + MgATP \overset{Na^+}{\underset{}{\rightleftharpoons}} E_1 \sim P + MgADP \tag{1}$$

$$E_1 \sim P \overset{Mg^{2+}}{\underset{}{\rightleftharpoons}} E_2 - P \tag{2}$$

$$E_2 - P \overset{K^+}{\longrightarrow} E_2 + Pi \tag{3}$$

$$E_2 \rightleftharpoons E_1 \tag{4}$$

It has never seemed likely that the hydrolysis of ATP takes place in a single step, and the simplest form generally accepted may consist of Eqs. (1) and (3) which are considered to reflect the phosphatase activity of the ATPase system. Equation (2) was postulated to be the step at which the Na, K, Mg-ATPase inhibitors activating the exchange reaction act, and Eq. (4) to be the step blocked by lowering temperature.

3.   Pump Model.

Many models have been proposed to reconcile the reaction scheme of
the ATPase with linked movement of Na^+ and K^+ across the membrane.
Although there are various types of models, such as a pump driven by un-
dulating peptide chains undergoing conformational change (52), a hole surrounded by charged walls of component enzymes catalyzing the ATPase
reaction (45), a membrane with sequential oxidation-reduction of sulfhydryl
groups induced by ATP (53), etc. (3, 54-56), they would be merely guide-
lines to help to suggest the experiments necessary to uncover new facts
about ion transport phenomena (52).   In general, they seem to agree that
the sequence of pumping action may be divided into two main steps like the
sequence of the ATPase chemical reaction: (a) phosphorylated intermedi-
ate(s) is formed and Na^+ ions move outward; (b) Pi is released inside the
cell and K^+ ions move inward.

4.   Molecular Weight.

The molecular weights of Na, K, Mg-ATPases of human red cell ghost,
microsome of guinea pig kidney cortex, and crayfish nerve cord were
estimated by the technique of radiation inactivation, to be approximately
250, 000 (57).   Mizuno et al. (58) determined the molecular weight of the
ATPase of pig brain microsomes by using a Sepharose 4B column and
sucrose gradient centrifugation.   They found the minimum particle size to
be approximately 500, 000 and suggested that the particle consisted of subunits, one of which possessed a molecular weight of 300, 000 as estimated
by radiation inactivation technique (59).   This suggestion may explain the
rather larger molecular weight of beef brain ATPase (670, 000) reported
by Uesugi et al. (60), using an agarose column.

The relation of active ion transport with Na, K, Mg-ATPase and the
characteristics of the enzyme were outlined above.   In order to learn in
more detail the underlying mechanism, in the following sections we
see the interaction between the membrane components to be organized
both to form the specific membranous structure and to carry out the
biochemical activity.

III.   MODE OF COMBINATION BETWEEN PROTEIN AND LIPID

According to the conventional trilamellar model of the membrane a
bimolecular leaflet of phospholipid-cholesterol mixture, with polar
ends facing outward, is covered with unfolded protein molecules.   The
interaction between the lipid and protein components is of an electrostatic
nature through polar terminals of phospholipid and the polar residues of
protein. The universality of the classic trilamellar model of the membrane
(61-63), however, has recently been increasingly questioned and challenged

through various lines of investigation, such as electron microscopy (64-73), biochemical study (74), spectral analysis (75-81), and theoretical treatment (82-85). The mode of combination between the membrane protein and lipid is being carefully reconsidered. The study on the relation of membrane ATPase with lipid, which is discussed later, also indicated that the conventional model may not explain the lipid action on the membrane protein to induce its normal catalytic and ion transport functions.

There are several reasons to believe that the manner of combination of lipid with protein in the membranous structure of the cell, at least in some cases, is hydrophobic in nature (74, 76, 78, 81, 86-92). Optical analysis of several different types of membranes such as red cell ghost, mitochondrion, myelin, chloroplast, plasma membrane of ascites tumor cell, and bacterial membrane, suggested that a large portion of the membrane protein is in α-helix. This proportion of highly ordered conformation may not permit the arrangement of protein in paucimolecular membrane. Furthermore, the common feature of a wide variety of membrane types in an optical rotatory dispersion spectrum suggested that the α-helix of the membrane protein was located in a hydrophobic environment (81, 91, 93). In a nuclear magnetic reasonance study (94) applied to human red cell ghost, the inhibition of signal due to $(CH_2)_n$ protons of the hydrocarbon chain was explained as nonpolar amino acids of membrane protein interacting with the hydrocarbon chain of lipid, and increasing magnetic dipole interactions between adjacent CH_2 groups. When the membrane was unfolded by trifloroacetic acid, many new peaks of amino acid were apparent and the $(CH_2)_n$ signal was intensified, which indicated that the hydrocarbon chains of the membrane lipid were partially interlocked with a portion of protein.

The easy accessiblity of phospholipase A (95-99), C (100-104), and D (105) to membrane lipid of erythrocyte ghost and other membranous structures of the cell suggested not only that membrane lipid was in contact with the bulk aqueous phase instead of being buried under a continuous protein sheet, but also that the polar and ionic ends of phospholipid were on the outer surface of the membrane. After a major portion of phospholipid of the membrane had been removed by phospholipase treatment, the membranes remained seemingly intact in phase microscopy as well as electron microscopy, and the average conformation of membrane protein was not changed when examined by circular dichroism measurement (101, 104). These facts would indicate that electrostatic interaction between the phospholipid and the protein of the membrane is not the main force which sustains the integrity of the membrane nor determines the conformation of the membrane protein.

From the hydrogen ion titration of the membrane of Halobacterium in various ionic strengths, Brown (106) suggested that the hydrogen bonds

between basic groups of the membrane protein and phosphate groups of
the membrane lipid played merely a secondary role to hydrophobic association of the nonpolar side chains since the membrane dissolved to give a
stable lipoprotein solution when the hydrogen bonds were disrupted, as
detemined by titration. The envelopes (membrane plus cell wall) of
Halobacterium were also tested by the techniques of dialysis, lipid extraction, solubilization, amino acid analysis, etc., by McClare (107), who
concluded that the lipids were bound hydrophobically to one group of
proteins and by polar bonds involving Mg^{2+} to another, and that differently
bound lipids seemed to be associated in the same complex, which was a
repeating membrane unit containing only one layer of lipid.

From their work, mainly on electron transport and other enzyme
systems of whole mitrochondria and of its component complexes, Green
and his collegues (74) showed the essentiality of lipid in the biochemical
and biophysical functions of mitochondrion. According to their theory,
the binding between adjacent repeating units is protein-protein in nature,
and lipid controls the kind and number of associations between repeating
units but is not involved in the bonding itself. Phospholipid enables the
lipid-free complexes of the enzyme system to align in a molecular film,
in which form the complexes are readily accessible to solute molecules of
the reaction medium. The charged group of phospholipids associated with
repeating units is fully exposed to the aqueous medium; the association is
through nonelectrostatic force.

Although the molecular architecture of the membrane proposed by
several authors mentioned above to replace the classical paucimolecular
model differs in detail, they agree that the binding manner of lipid to
protein is in all likelihood nonpolar interaction. Na, K, Mg-ATPase and
K^+, Mg^{2+}-activated phosphatase (K, Mg-phosphatase) of the cell membrane
are lipoprotein in nature and the enzyme proteins have to be attached to
appropriate lipids in order to fully carry out their catalytic functions.
The interaction of the enzyme protein with lipid was again interpreted
to be, at least partly, nonpolar in nature, from the study on the activation
of those two enzymes by a series of synthetic and natural lipids. The manner of interaction with lipid and various aspects of the enzymatic alteration
of the membrane protein induced by the combination with lipids, is discussed in detail in the next section.

IV. EFFECTS OF LIPID ON ATPase AND PHOSPHATASE

A. Lipid and Enzyme Activity of Membrane

The cell membrane is composed mainly of protein, phospholipid, and
small amounts of polysaccharide. These three components are integrated

not merely to form a stable boundary structure to mechanically protect the protoplasm from its surroundings, but also to let the protein component carry out physicochemical activity by providing it with proper molecular conformation and with specific association between adjacent molecules. Even the minor component, saccharide, seems to be associated with the biochemical activity of the membrane and once the arrangement is disturbed the biochemical activity of the membrane is lost (though the manner of interaction is entirely unknown), since culture filtrate of <u>Vibrio cholerae</u> which contained mucinase selectively destroyed Na, K, Mg-ATPase activity (108). This thesis would also be justified by the fact that the contact of organic solvents or detergents, even at very low temperatures, with the ATPase or membrane preparation eliminates most of the ATPase activity (109). This was further substantiated by the data showing that by incubating with phospholipases red cell ghosts lost the ouabian-sensitive ATPase, and the activity was restored by the addition of phosphatidyl serine to the reaction mixture for ATPase determination (96, 98, 100, 101, 104). Experiments of these types have, in general, intrinsic shortcomings in that the reaction products resulting from the incubation with hydrolyzing enzymes may act as inhibitors or at least as modifiers of the structure of enzyme protein.

In order to conclusively show the effects of phospholipid on an enzyme composed of protein and phospholipid, first the enzyme has to be purified to a great extent and second the enzyme activity should be demonstrated to be eliminated by the removal of the lipid and to be restored by the recombination of lipid with the enzyme protein. No perfect report of this type has been presented with regard to the ATPase, mainly because of the difficulty in the purification of the complicated enzyme system and the incompatibility of the first and second conditions. However, some of our experiments partially satisfy the conditions given above. We now discuss the functional relationship between membrane ATPase and lipid, mainly based on those data.

B.   Treatment of Membrane With Deoxycholate

In the effort to remove the lipid component from the particulate fractions of tissues (while preserving their enzyme activity), various types of agents have been used such as organic solvents (tert-amyl alcohol, butanol, acetone-water mixture), detergents (deoxycholate, cholate, Triton X-100 Tween 40), phospholipase A plus vigorous washing, and salts with organic solvents (110). Triton X-100, butanol, and water-acetone inactivated the membrane ATPase, and are not usable for the present purpose (111). The combination of deoxycholate treatment and ammonium sulfate fractionation was used with some success to separate the phospholipid component from the protein component of membrane or microsomes obtained from brains of rat, cow, rabbit, cat, guinea pig, pig, and kidneys of cow, pig, rabbit, and rat.

The cerebral membrane fraction was usually obtained between mitochondrial and microsomal fractions, as "heavy" microsomal fractions either by conventional differential centrifugation (112) or gradient centrifugation with Ficoll (111, 113). The solubilized fraction obtained after the treatment of the membrane fraction with deoxycholate at $0°*$ was subjected to ammonium sulfate fractionation. The concentration of sodium deoxycholate was less than 0.5% and the weight ratio of protein to the detergent was kept at 1:1. Since most of the lipid phosphate in the solubilized fraction was removed by ammonium sulfate fractionation, the membrane had presumably been separated into the phospholipid and protein moieties by exposure to deoxycholate (Table 1.). This idea is supported by the observation of the spectral change in nuclear magnetic reasonance of red cell ghosts after deoxycholate treatment. The data were interpreted to show that deoxycholate liberated the lipid component from the protein component of the membrane and then formed mixed micelles with the liberated lipid (94, 114).

TABLE 1

Activity and Lipid Phosphate of Deoxycholate-Treated ATPase
in Supernatant after Ammonium Sulfate Fractionation

	ATPase activity			
Ammonium sulfate concentration % saturation	Crude animal lecithin[a] (1 mg/tube)	Pi liberated (μmoles/mg protein)	Protein (mg/ml) of enzyme)	Lipid phosphate (μmoles/ml of enzyme)
0	−	1.15	2.98	1.00
	+	1.14		
10	−	1.19	1.04	0.35
	+	1.22		
20	−	0.55	0.79	0.09
	+	1.65		
30	−	0.43	0.54	0.03
	+	1.54		

[a]Lipids used were dispersed in water or tris buffer by means of either a Potter-Elvehjem type homogenizer made of Teflon and glass or sonication.

*All reference to temperature is in degrees centigrade.

As the concentration of ammonium sulfate was increased, the activity of the ATPase remaining in the supernatant fraction decreased and was roughly parallel to the loss of lipid phosphate from the fraction (Table 1; the ATPase shown in the tables was prepared from a membrane fraction of the bovine cerebral cortex). It may well be that the ATPase activity was lost due to the removal of phospholipid from the lipoprotein complex of the membrane.

C. Recovery of ATPase Activity by Phospholipids

If the loss of the enzyme activity had been due to the removal of phospholipid of the lipoprotein complex of the enzyme, then the adding back of phospholipid would have restored the activity. This was the case. Table 2 shows the activation of Na, K, Mg-ATPase by crude animal lecithin (112). Since the amount of ATP hydrolyzed by the enzyme preparation did not differ before and after the addition of Na^+ plus K^+, this small ATPase activity given in the absence of the added phospholipid would not be of Na, K, Mg-ATPase. The addition of crude animal lecithin did not induce any change in this small ATPase activity. When crude lecithin was added along with Na^+ and K^+, however, the activity increased 2.5-fold. Consequently, only Na, K, Mg-ATPase is activated by crude lecithin; an inactive form of the ATPase is changed into an active form by the combination with phospholipid. That Na, K, Mg-ATPase alone is activated by phospholipid is

TABLE 2

Effects of Na^+, K^+, Mg^{2+}, and Crude Animal Lecithin on ATPase

NaCl (mM)	KCl (mM)	MgCl$_2$ (mM)	Crude animal lecithin	Pi liberated (μmoles/tube)
100	10	3	−	0.12
			+	0.33
110	0	3	−	0.12
			+	0.15
0	110	3	−	0.12
			+	0.13
0	0	3	−	0.13
			+	0.13

substantiated by the fact that the ATP hydrolysis depending on the presence of crude lecithin, Na^+ and K^+ was inhibited by 6×10^{-6} M ouabain to 50% (Table 3) (112). In the absence of Na^+ plus K^+, even 1×10^{-3} M ouabain inhibited the hydrolysis to less than 10%.

The crude animal lecithin preparations used were found by thin layer chromatography to contain at least five different components: licithin, phosphatidyl serine, phosphatidyl ethanolamine, neutral lipid, and lyso-compounds. A question naturally arose as to which component(s) of these lipid preparations activated Na, K, Mg-ATPase. The effects of the pure preparations of various natural phospholipids from a variety of sources were tested and are listed in Table 4 (115). Cardiolipin, lecithin (egg and dipalmitoyl), phosphatidyl ethanolamine (egg and dipalmitoyl) were ineffective in activation.

D. Recovery of Phosphatase Activity by Phospholipids

There were at least three different types of phosphatases in the membrane fraction of bovine cerebral cortex (116): (a) a phosphatase with no metal requirement, pH optimum 4. 6 (acid phosphatase); (b) one requiring Mg^{2+}, pH optimum 6. 2 (Mg-phosphatase); (c) one activated by K^+ plus

TABLE 3

Inhibition of ATPase by Ouabain

Addition	Ouabain (M)	Na, K, Mg-ATPase (%)
Crude animal lecithin (250μ g/ml)	0	100[a]
	1×10^{-6}	63
	1×10^{-5}	15
	1×10^{-4}	4
Didecyl phosphate (0. 6 mM)	0	88
	5×10^{-5}	10
Monotetradecyl phosphate (0. 4 mM)	0	30
	5×10^{-5}	13

[a]This value is arbitrarily set at 100% for the purpose of comparison. The inhibition of Mg-ATPase by 1×10^{-3} M ouabain was less than 10%.

TABLE 4

Activation of Na, K, Mg-ATPase by Natural Phospholipids

Phospholipids	ATPase activation (%)
Crude animal lecithin	100[a]
Phosphatidyl serine (bovine brain)	100
Phosphatidic acid (egg)	97
Lysophosphatidyl ethanolamine	96
Phosphatidyl serine (egg)	65
1-Palmitoyl lysolecithin	41
Phosphatidyl inositol (soybean)	28
Lecithin (bovine brain)	20

[a]The value is set arbitrarily at 100% for comparison. The concentration of phospholipids is 250 μg/ml.

Mg^{2+}, pH optimum 8.6 (K, Mg-phosphatase). In the supernatant obtained after centrifugation of the deoxycholate-treated membrane fraction at 140,000 x g for 45 min, all three phosphatase activities were detected along with Na, K, Mg-ATPase and Mg-ATPase. Although two ATPases were partially separable by ammonium sulfate fractionation, Na, K, Mg-ATPase and K, Mg-phosphatase were inseparable.

Of the three phosphatases, only K, Mg-phosphatase activity was affected by removal and recombination of the lipid component of the membrane. Like Na, K, Mg-ATPase activity, K, Mg-phosphatase activity of the membrane fraction was decreased by deoxycholate treatment but not completely lost, which was dissimilar to the case of the ATPase. The combination of the lipid resulted in stimulation of K, Mg-phosphatase (116). Crude animal lecithin, for instance, increased the phosphatase activity approximately 70% from the original activity obtained in the absence of exogenous phospholipid. The range of chemical structure of the required phospholipids was wider in the case of phosphatase than it was in the case of the ATPase (Table 5).

TABLE 5

Activation of K, Mg-Phosphatase by Natural Phospholipids

Phospholipids	Phosphatase activation (%)
Crude animal lecithin	100[a]
Phosphatidyl serine (egg)	166
Lecithin (bovine brain)	84
Lysophosphatidyl ethanolamine	71
Phosphatidic acid (egg)	67
1-Palmitoyl lysolecithin	46
Licithin (egg)	43
Dipalmitoyl lecithin	37
Phosphatidyl inositol (soybean)	35
Phosphatidyl serine (bovine brain)	28
Phosphatidyl ethanolamine (egg)	19

[a]This value is set arbitrarily at 100% for the purpose of comparison. The lipid used was 250 μg/ml of reaction mixture.

Contrary to phospholipid, neutral lipid inhibited the phosphatase activity. At the concentration of 250 μg/ml, both oleic and linoleic acids inhibited the activity almost completely and monolein reduced it 30%, diolein and triolein effects being smaller. Nonanoic acid, a saturated fatty acid, also strongly inhibited the activity at high concentrations.

E. Molecular Structure in Lipid Essential to Activation

1. ATPase.
Although a variety of phospholipids, as described above, were shown to activate the ATPase of the membrane fractions of brain and kidney of cattle, cat, rabbit, and rat, all four sectors composing phospholipid (amino alcohol

or choline, phosphate, fatty acid, and glycerol) would not appear to be essential to activation. Because licithin and phosphatidyl ethanolamine activated the enzyme only slightly, and phosphatidic acid was highly effective in the activation, amino alcohol or choline moiety did not seem essential to the activation. A phosphate group, or a negatively charged sector, however, may be required, since most of mono-, di-, and tri-glycerides tested were ineffective (117). This view is substantiated by the fact that acidic phospholipid, like phosphatidyl serine or phosphatidyl inositol, activated the ATPase but neutral phospholipid, such as phosphatidyl ethanolamine or lecithin, did not. A fatty acyl moiety as well as a phosphate group would be needed for activation, inasmuch as lysolecithin and lysophosphatidyl ethanolamine strongly activated the enzyme (117) but glycerophosphate was ineffective. Lecithin preparation from beef brain activated the enzyme to some extent, while dipalmitoyl lecithin did not. These facts may indicate that the effectiveness in activation partly depends on the number of fatty acyl moieties and degree of their saturation and that a hydrocarbon chain rather than a carboxyl group moiety plays an important role. It is not clear, however, whether the significance of the double bond in ATPase activiation is to lower the hydrophobic character of phospholipid or to determine the proper conformation of phospholipid to facilitate the combination with the enzyme protein. The compound needed for the ATPase activation, therefore, should consist of at least a phosphate group and a fatty acyl residue or a hydrocarbon chain. While the importance of glyceryl residue is not known from the cited data, monopalmitin, dipalmitin, diolein, triolein, trilinolein, and α-glycerophosphate were ineffective (115).

The above conclusion regarding the molecular structure essential to activation is supported by the fact that alkyl phosphates were markedly effective in activation of the ATPase (Table 6) (117), and the conclusion may be modified as follows: (a) Instead of a fatty acid residue a fatty alcohol residue is sufficient, and the hydrocarbon chain need not be unsaturated if the chain length is appropriate. (b) The number of carbon atoms in the hydrocarbon chain decides the hydrophilicity-hydrophobicity balance as well as the binding force and the conformation appropriate for combination with the ATPase protein, because both mono- and dialkyl phosphates possessing ten-carhon chains were most effective, and mono- and dioctadecyl phosphate showed little or no effects on either the ATPase or the phosphatase. Thus the chain length determines the effectiveness of activation. (c) A phosphate residue can be replaced by a sulfate residue or a carboxyl radical, although the efficacy of alkyl sulfate and fatty acid was weaker and more limited than that of alkyl phosphates. The range of activating concentration and the degree of activation by dodecyl sulfate were one-fifth and one-half, respectively, of those by monododecyl phosphate. At concentrations higher than 0.9 mM the alkyl sulfate inhibited completely Na, K, Mg-ATPase, Mg-ATPase, K, Mg-phosphatase,

TABLE 6

Activation of Na, K, Mg-ATPase by Di- and Monoalkyl
Phosphate and Related Compounds

Compounds	Optimum concentration (mM)	Activation of ATPase (%)
Crude animal lecithin	250 μg	100[a]
Monohexyl phosphate	8. 2	29
Monoctyl phosphate	4. 8	49
Monodecyl phosphate	2	100
Monododecyl phosphate	0. 2	45
Monotetradecyl phosphate	0. 43	31
Monoctadecyl phosphate	0. 4	17
Dihexyl phosphate	0. 48	73
Dinonyl phosphate	0. 18	100
Didecyl phosphate	0. 85	120
Didodecyl phosphate	1. 1	55
Dodecyl sulfate	0. 05	58
Nonanoic acid	9	47[b]
Decyl alcohol	2	41
Dodecyl alcohol	0. 9	23
Linoleyl alcohol	0. 5	16[b]
Linolenyl alcohol	0. 5	12[b]
Monolein	0. 7	22[b]

[a]This value is set arbitrarily at 100% for the purpose of comparison.

[b]The value is not maximum, but is the highest value obtained in the tested
range.

Mg-phosphatase, and acid phosphatase. Though a 3- to 5-fold higher concentration of nonanoic acid was needed to obtain the same degree of the
ATPase activation as given by monotyl phosphate, the nonanoate effects
were similar to those of monoctyl phosphate. The tridecanoate effects
were almost identical to those of monotetradecyl phosphate. (d) A phosphate group cannot be replaced by amine residue without losing the activating effect, since dodecyl and octadecyl amines were inactive.

2. Phosphatase.

    In the previous section we saw that a negatively charged group with a
hydrocarbon chain or two is essential to activate Na, K, Mg-ATPase.
This is also the case for K, Mg-phosphatase of the membrane, but the
structure should contain two chains. So far as phospholipid is concerned,
however, the essential structure needed for the phosphatase stimulation
does not differ very much from that for the ATPase activation (Table 5).
There are two differences worth mentioning: (a) Neutralization of the
phosphate residue by a positively charged moiety attaching to the residue
does not impair the ability of the lipid to stimulate the phosphatase, since
licithin and phosphatidyl ethanolamine were active. (b) The presence of a
double bond in the hydrocarbon chain of phospholipid is not required to
stimulate the phosphatase, since synthetic dipalmitoyl lecithin, as well as
natural lecithin, was effective.

    The data obtained with alkyl phosphates (Table 7) suggest the following:
(a) Two hydrocarbon chains in an alkly phosphate molecule are required to
stimulate the phosphatase, and monoalkyl phosphate is generally inhibitory.
Monoalkyl phosphate, as the chain elongated, decreased the inhibitory
action, and the alkyl phosphates with 12 and 14 carbons in the hydrocarbon
chain were slightly stimulating at low concentrations. Polystyrene phosphate, in which phosphate residues are attached to a rigid polystyrene
framework, showed little inhibition. (b) Dialkyl phosphate with ten carbon
atoms in each of its hydrocarbon chains was most effective in the phosphatase stimulation; and showed a very similar pattern of stimilation (the
concentration of decyl phosphate versus the activity of the phosphatase) to
that of natural phospholipid. (c) A phosphate residue as a negatively charged
sector in alkyl phosphate is more critical to stimulate phosphatase than
it is to activate the ATPase, because dodecyl sulfate merely inhibited
the phosphatase and trideconoate was only half as active as monododecyl
phosphate. (d) When an amino radical replaces a phosphate group the
resulting compound is strongly inhibitory, as shown in the cases of dodecyl
and octadecyl amine. (e) Determining molecular conformation and hydrophilic tendency, a double bond is important in determining the action of
alcohols on the phosphatase. At a concentration of 0.5 mM, oleyl alcohol
inhibited the phosphatase very slightly, and linoleyl and linolenyl alcohols
were 10- and 13-fold more inhibitory, respectively, than oleyl alcohol.

TABLE 7

Stimulation of K, Mg-Phosphatase by Mono- and Dialkyl Phosphates
and Related Compounds

Compounds	Optimum concentration (mM)	Stimulation of phosphatase (%)
Crude animal lecithin	250 μg	100[a]
Monododecyl phosphate	0.2	53
Monotetradecyl phosphate	0.43	23
Dihexyl phosphate	0.24	45
Dinonyl phosphate	0.18	75
Didecyl phosphate	0.85	100
Didodecyl phosphate	0.15	23
Tridecanoic acid	0.3	28

[a]The value is set arbitrarily at 100% for the purpose of comparison.

When a double bond is present in fatty acid, it is highly inhibitory. At the concentration of 0.9 mM oleic acid inhibited the phosphatase over 90% and the inhibition of linoleic acid was almost complete, whereas tridecanoic acid was slightly activating. (f) As the number of hydrocarbon chains increases, which also affects the hydrophobic character and decides the molecular shape, inhibitory effects on the phosphatase diminish. Monolein inhibited the enzyme to approximately two-thirds of the original activity at a concentration of 250 μg/ml while diolein inhibition was one-third of this value and triolein was virtually not inhibitory. Without a double bond, mono- and diglycerides show no effect on the enzyme activity.

3. Alcohols.

    Long chain fatty alcohols and monolein possess peculiar properties somewhat resembling those of monoalkyl phosphates, which activate the ATPase while inhibiting the phosphatase. Activating effects of fatty alcohols, either with or without double bonds, are generally much weaker than monoalkyl phosphates, and oleyl alcohol shows no significant effect on either enzyme whereas other alcohols with more than one double bond activate the ATPase and inhibit the phosphatase (Table 6). It is interesting to note that the structures of the alcohols and monolein, with their hydrophilic residue(s) located at one end of a long hydrophobic chain, are basically similar to that of monoalkyl phosphates and, as mentioned above, the action of the compound would be determined mainly by the properties of their hydrocarbon chain. This would explain the resemblance in the effects of those three types of seemingly unreleated compounds. The alternative explanation may be that the alcohols and monolein combine with endogenous activator(s) and inhibitor(s) to make them more accessible to the enzymes, or the binding of the compounds to the enzyme make endogenous activator(s) and inhibitor(s) more accessible to the enzymes.

4. Comparative Study.

Our conclusion on the molecular structure in phospholipid essential to activate the enzymes is based on the experiments with an enzyme preparation from bovine cerebral cortex. Is the essential structure needed for activation limited to this particular tissue of this particular species? In order to answer this question rats, cats, rabbits, hogs, and sheep, and livers, kidneys, hearts, as well as brains were tested with regard to the lipid activation of the ATPase and the phosphatase of the membrane fraction.

The ATPase and the phosphatase prepared form tissues other than brain were, in general, activated by crude lecithin, and the activation and inhibition patterns of brain enzymes were similar regardless of species of animal.

Monoalkyl phosphates tended to activate the kidney ATPase more than the brain enzymes, including bovine kidney, and was less inhibitory to the phosphatase of bovine kidney. Dialkyl phosphates showed no significant activation of the ATPase and the phosphatase of heart. Dialkyl phospates with shorter alkyl chains than the length of ten carbons tended to activate the kidney enzymes more than the brain enzymes regardless of species of animal. The phosphates with longer chains than nine carbons showed a similar degree of activation of the ATPase of all tissues and of all species tested, whereas the same phosphates tended to inhibit the phosphatase of kidney at lower concentrations than expected from the results with the brain enzyme.

Although there were small but apparent differences in the response pattern of the membrane enzymes to various alkyl phosphates between bovine cerebral cortex and other tissues of the other animal species, the general activation and inhibition pattern remained essentially the same. Consequently, the molecular structure in phosphlipid needed for activation of the membrane enzymes would be common among the tissues and the species, at least in the tested range.

Those small differences observed with various tissues and species may be due to the structural differences in the enzyme proteins inducing the variations in affinity with the added artificial lipids, or due to the differences in susceptibility of the membranes to deoxycholate treatment, which removes phospholipids from the membrane, or due to both. Though the first explanation is more attractive, we are not yet at the stage to conclude its validity.

F.   Mechanism of Action of Lipid

1.   Manner of Combination.

As for the combining forces of the enzyme protein with either lipid or alkyl phosphates, three types may be considered: electrostatic force, hydrogen bonding, and van der Waals force. Since, as described above, a hydroxyl group can replace a negatively charged group at an end of long-chain alkyl compounds without completely losing activating effects on either the ATPase or the phosphatase, electrostatic force may not play a central role in combination. Inasmuch as the alkyl chain is absolutely needed for the activation and the alkyl compound shows the highest activation effects when it possesses ten carbon atoms in a single chain, the combination seems to be controlled by the hydrocarbon chain; the binding of lipid to the enzyme protein may be attributed mainly to van der Waals force. This view is from an analogy to the study on the combination mode between enzymes and substrates (118).

2.   Protection from Thermal Inactivation.

When the ATPase preparation was incubated at $25°$ for 30 min prior to the activity determination, the activity of Na, K, Mg-ATPase was decreased to 55% of the original (112). If the preincubation was carried out after the enzyme preparation had been bound to phospholipid, however, the enzyme activity remained at almost the same level as the original. Similar results were also obtained with K, Mg-phosphatase. Combination with the lipid increased the resistance of the enzyme protein against thermal inactivation. This may be explained by either the stabilization of protein conformation by binding of lipid or the prevention of the enzyme protein from random aggregation which hinders the substrates' access to the active sides of the enzyme.

3. Relationship of Phospholipid to SH Group.

It is well documented that SH groups are essential to the activity of the ATPase of the membrane, although the group does not seem directly involved in the reaction as a phosphorylated site of the enzyme protein (2, 5). This fact was also demonstrated with the deoxycholate-treated ATPase, which was extremely sensitive to SH inhibitors such as HgCl$_2$, p-chloromercury benzoate, and N-ethyl maleimide (112). The inhibitory action was eliminated by the addition of sulfhydryl reagents. The fact that the deoxycholate-ATPase also lost its activity in the presence of either Cd^{2+} or arsenite plus 2,3-dimercaptopropanol at low concentrations (112) suggests that the sulfyhdryl group involved in the activity of the ATPase is dithiol in nature, since it has been concluded that the uncoupling effects by those reagents show the involvement of dithiols in the oxidative phosphorylation process of rat liver mitochondria (119, 120).

The sulfhydryl groups involved in the ATPase activity were protected by the combination with phospholipid against the attack of an SH inhibitor, although the detailed mechanism has yet to be defined, for instance, as to whether the protective action of phospholipid is due to the conformational change induced by the lipid binding or due simply to the steric hindrance of the attached lipid. The preincubation of the enzyme preparation with N-ethyl maleimide prior to the activity determination suppressed the activity to 15% of the original level, but when the crude lecithin was added to the preparation before the preincubation with the inhibitor, the activity remained at the 50% level (112).

The sulfhydryl group involved in the activity of the ATPase may not be oxidized during the purification procedure, nor may the group involved in the combination with the lipid be affected, since the addition to either the reaction mixture or the enzyme preparation of SH-reagents like glutathione, cysteine, and 2-mercaptoethanol increased the ATPase activity 20% at best, and the sulfhydryl reagents did not decrease the rate of inactivation of the prepared ATPase.

4. Ratio of Lipid to Enzyme Protein.

When the concentration of the enzyme protein was maintained constant and the activity was determined in the presence of varied amounts of didecyl phosphate, there was an optimum concentration for the alkyl phosphate (121). As the concentration of the enzyme protein was changed, the optimum concentration was altered proportionally; the ratio of the amount of the alkyl phosphate for optimum activation to the amount of the protein remained constant. A similar relation was observed in the case of the phosphatase. Thus, the arrangement or the conformation of the enzyme protein molecules may be most suitable for enzymatic activity when lipid is bound to the protein in a proper ratio. Further, it is unlikely that the combination of lipid with the enzyme protein is reversible under the incubation conditions for activity determination.

5.   Km of ATPase.

Although the activity of the ATPase varied according to the amount of added phospholipid, the Km value calculated from double-reciprocal plot remained unchanged, at least at low concentrations of the lipid (121). The addition of more phospholipid seemed to simply mean the addition of more enzyme preparation. The Km value of the ATPase obtained by deoxycholate treatment of a membrane fraction of bovine cerebral cortex was identical to that of the ATPase of the untreated fraction itself, and did not change according to the species of the lipids used to activate the enzyme. It is just as if the combination of phospholipid with the enzyme protein in a certain ratio caused a simple change from an inactive form into a fully active form and no further change, and as if there were only a single active form of the ATPase, which was unrelated to the structures of individual molecules of the lipid activating the enzyme.

6.   State of Lipid.

Either the molecularly dispersed form of lipid or the micellar form is effective to activate the ATPase and the phosphatase. When dispersed in an aqueous solution, natural phospholipid is believed to be in a micellar state even at extreme dilution instead of assuming the state of molecular dispersion because of its strong hydrophobic character (74). Since bovine phosphatidyl serine activated the ATPase and the phosphatase at any concentration, a micellar state would be an effective form of the phospholipid. Similarly, 1-palmitoyl lysolecithin showed no lag phases in the activation curves (activities versus lipid concentrations) (116) but, contrary to the case of natural phosphatidyl serine, the lysolecithin is present in both a micellar state and molecular dispersion in an aqueous solution. Indeed, the highest activation of the enzymes was obtained at the concentration below its critical micelle concentration. The critical micelle concentration of the lysolecithin is approximately 3×10^{-5} M in the reaction mixture for either ATPase or phosphatase determination, when determined by the capillary rise method (121). A similar relation between the critical micellar concentration and the optimum activation concentration was obtained with both mono- and didecyl phosphates (117). That those two alkylphosphates, especially mono-compound, easily passed through a usual cellophane dializing tube further supports the view that those alkyl phosphates are in an equilibrium between micellar form and molecular dispersion in an aqueous solution. Though the arrangement of lipid molecules after they have been bound to the enzyme protein is not yet known, the state of lipid dispersed in the reaction mixture, whether it is in the form of molecular dispersion or in micellar form, is not important for the activation of the membrane ATPase and phosphatase.

7. pH Optimum Shift.

Another change caused by deoxycholate treatment and combination with the lipid is the pH optimum shift. The pH optima of Na, K, Mg-ATPase of cerebral membrane fractions prepared from Japanese cattle and rabbit decreased by 0. 8 pH units, whereas those from American cattle, cat, guinea pig, and rat showed no shift (115). The pH optimum shift was uniformly from pH 8. 6 to 7. 4 with the phosphatase of the cerebral membrane fractions from American cattle, rabbit, cat, guinea pig, and rat, except for that from Japanese beef cattle which gave the shift from pH 7. 8 to 7. 4.

The shift depended on none of the following factors: preparations of deoxycholate, phospholipid, and substrate; concentrations of deoxycholate; and ratio of deoxycholate to membrane protein.

8.   Binding Sites of Lipid.

Because even a higher concentration of crude lecithin did not change the ouabain inhibition, the binding site of the lipid would not be related to the ouabain binding site on the enzyme protein. Similarly, since the activation patterns obtained by varied concentrations and varied ratios of Na$^+$ and K$^+$ were not influenced by changing the amount of the added phospholipid, the binding sites of lipid and the ions may not be related either (112).

V.   RELATION BETWEEN ATPase AND PHOSPHATASE

A.   Particulate Enzyme Preparation

There has been much controversy over the identity of Mg, K-phosphatase of the membrane with respect to its relation with Ka, K, Mg-ATPase, and it seems now to be a general trend to regard the phosphatase activity as a reflection of a partial reaction catalyzed by the ATPase system, although its detailed relation, in terms of its chemical reaction scheme and ion transfer mechanism, is not yet fully defined. Because a number of similarities between those two enzymes were observed in the effectiveness of various inhibiting agents (ouabain, p-chloromercury benzoate, N-ethylmaleimide, diisopropylfluorophosphate, and heating), in the increase of specific activities during purification and in the rise of activities during development of the brains of a variety of animals, and because there was mutual inhibition by substrates of those two enzymes, several investigators considered the phosphatase activity a part of the reactions carried out by the ATPase (122-126), though the relation may not be as simple as has been thought (127). On the other hand, Albers and his co-workers (128, 129) regarded the phosphatase as a separate enzyme from the ATPase system, based on the differences in Km for Mg^{2+}, in the activation and inhibition pattern by various cations (K^+, Li^+, Rb^+, Cs^+, and Ca^{2+}), in the effective concentration of ouabain, in pH optimum, and in the distribution pattern among subcellular fractions. In the early experiments p-nitrophenyl phosphate, one of the commonest substrates for phosphomonoesterase, was used as substrate for determination of K, Mg-phosphatase. When carbamylphosphate and acetyl phosphate, more suitable substrates in this case than p-nitrophenyl phosphate, were employed as substrates, however, some of the above-mentioned discrepancies disappeared and, in addition, Km's (for Mg^{2+}, K^+, Rb^+, NH_4^+, and Li^+) and Ki's (for ouabain, Ca^{2+}, Hg^{2+}, and F^-) were found similar, and both enzyme activities possessed similar activation energy (130-133).

B.   Deoxycholate-treated Preparation

The enzyme preparations used for those studies were either subcellular particulates or insoluble lipoprotein complexes. After deoxycholate treatment, which has removed approximately six-sevenths of phospholipid contained initially in the membrane fraction of the brain, the ATPase and the phosphatase displayed similarities and also differences, especially in the effects of lipids as described in the previous section.

The following facts observed with the deoxycholate-treated preparation support the view that the phosphatase is a part of the ATPase system (116): (a) The phosphatase activity of the deoxycholate-treated preparation was extremely thermolabile, and the time curve of the thermal inactivation of the enzyme activity was essentially identical with that of the ATPase activity. (b) Even under the protective action of phospholipid, the time curves of the thermal inactivation of both enzymes were also identical (116). (c) Sulfhydryl reagent inhibited the activities of both enzymes to the same extent (112). The parallelism in the inhibition by SH-reagents, however, may not be evidence supporting the theory that one enzyme is part of the other enzyme system, since the reagents lower in the same manner all of the activities of three distinctly independent phosphatases of membrane (acid, Mg^{2+}-activated, and K^+, Mg^{2+}-activated) (116). (d) Both enzymes of the doxycholate-treated preparation were inseparable by ammonium sulfate fractionation (115). (e) ^{32}P which has been incorporated from acetyl phosphate-^{32}P into the protein fraction of the enzyme preparation in the presence of K^+, Mg^{2+}, and crude lecithin was decreased to a similar degree by the addition of either nonradioactive ATP (plus Na^+) or acetyl phosphate. Both labelling of the enzyme protein fraction and lowering of the specific radioactivity by the addition of the substrates were almost completed within 2 sec at $25°$ (121). The reduction in the specific radioactivity seemed to be attributable to the dilution effect by formation of an intermediate common to the hydrolysis reactions of either ATP or acetyl phosphate. The possibility may be excluded that the radioactivity decrease upon the addition of cold ATP (plus

Na$^+$) is due to the inhibition of the intermediate formation of the phosphatase reaction.

As is easily seen from the discussion in Sec. IV, C-E, there are considerable differences between the ATPase and the phosphatase with respect to the response to lipid. Further, the specific activity ratio of the ATPase to the phosphatase does not remain constant during deoxycholate treatment and ammonium sulfate fractionation. Though the ratio was 2. 2:1 with the initial membrane fraction, it was 7. 9:1 with the deoxycholate preparation and was 4. 4:1 in the presence of crude egg lecithin (115).

Although it is tempting to draw a definite conclusion as to whether or not the two enzymes are separate entities, at the moment the precise relation of the phosphatase to the ATPase has yet to be fully understood. In order to explain all of the facts mentioned, however, it would be more reasonable to regard the phosphatase as a part of the ATPase system, making some assumptions. If K, Mg-phosphatase is considered to be a separate enzyme entity from Na, K, Mg-ATPase, the fact described in (e) regarding ^{32}P uptake from acetyl phosphate-^{32}P will be hard to explain. On the other hand, if the phosphatase is regarded as reflecting a paritial reaction catalyzed by the ATPase system, the mechanism of action of lipids on the phosphatase will have to be looked upon as different from that on the rest of the system. In order to account for their differences in lipid effects, it would have to be further assumed that the hydrolyzing reaction of a true substrate, which is an intermediate derived from reaction between ATP and the system, is not inhibited by the listed compounds, whereas the reaction of exogenous substrates for phosphatase activity is inhibited.

The inhibition by monoalkyl phosphate is at least partly due to competition between the phosphatase substrate and alkyl phosphate, since phosphate was released from monohexyl phosphate in the presence of either Mg^{2+} only or Mg^{2+} plus K$^+$ (117), and since monodecyl phosphate showed a pattern of competitive inhibition when added with p-nitrophenyl phosphate, a phosphatase substrate (121). It is not unreasonable, therefore, to assume that so far as monodecyl phosphate is concerned the lipid inhibition occurs at a reaction step before an intermediate common to the ATPase reaction is formed and thus hydrolysis of exogenous substrates for the phosphate is inhibited although hydrolysis of endogenous substrate formed by ATPase reaction is not affected. This assumption is supported by the fact that monodecyl phosphate inhibited ^{32}P incorporation from labeled acetyl phosphate into protein fraction of the enzyme (121).

VI.   ATPase IN NERVOUS SYSTEM
There is a close association of Na, K, Mg-ATPase with the various aspects of the activity of the nervous system. The importance of the

enzyme in the system seems to be based mainly on the role of the enzyme in ion transport to maintain the normal conducting function by sustaining physiological ion distribution on both sides of the neuronal cell membrane.

A.   Localization of ATPase

Localization of the ATPase was determined by both biochemical and histochemical methods.   The results of both approaches have seemed to agree well and have supplemented each other.

1.   Tissue Level.

Na, K, Mg-ATPase has extremely wide distribution among animal tissues (134), and of these tissues intestine, skin epithelium, kidney, secretory organ, electric organ, and above all nervous system are known to be rich in the ATPase (135, 136).   Differences in the ATPase activity were noted between gray matter (highest in activity), white mater, various nerve fibers, retina and ganglions of cat (136).   In rhesus monkey brain, the highest activities were found in the cerebral cortex, cerebellar cortex, thalamus, and colliculi, and the lowest in white matter.   The specific activity of the enzyme in gray matter was regular in distribution with a twofold difference between the highest and the lowest (137).   Harmoney et al. (138) found that in the central nervous system of rabbits mesencephalon was most abundant in the ATPase activity, cerebellar cortex, cerebral cortex, and medulla followed, and lumbar spinal cord, cervical spinal cord, and corpus callosum were next, and optic chiasm was lowest; i. e., the nervous structures richer in nerve fibers have a lower Na, K, Mg-ATPase activity. No significant differences were observed in the ATPase activity between symmetrical regions of the cerebral cortex in the same animal.   Although some activity differences were noted among various regions of cortices, it is not clear whether or not the findings were significant, since only one animal was tested.

2.   Subcellular Level.

As the membrane of red cell has been known to possess the ATPase activity and detailed localization was confirmed by histochemical technique (139), much evidence is accumulating with regard to the subcellular locali-zation of the enzyme in the nervous system.   When Skou (1) reported Na, K, Mg-ATPase for the first time, the material used was microsomal fraction of crab nerve fibers, which was believed to contain the major part of the axonal membrane.   Later the ATPase was confirmed to be localized in the neuronal membranes of the vestibular nucleus and microsomal fractions obtained from the brain were found to contain glia and nerve-cell membranes (140).

Applying the microchemistry technique to rat somatosensory cortex, Lewin and Hess (141) showed that the peak of Na, K, Mg-ATPase activity was present in the lower part of layers I, IIIa, IIIb, and Va and in the upper part of layer VIa, and that the distribution pattern resembled that of acetylcholine esterase and was also consistent with the distribution pattern of finely divided plexi of dendrites and axons, where neuronal plasma membranes were superabundant. Neurons were microdissected from surrounding glia in the lateral vestibular nucleus of the rabbit brain, and Na, K, Mg-ATPase activity was found to be confined to the neurons (140). Furthermore, Hess et al. (142) observed that Mg-ATPase of human astrocytomas was not stimulated further by the addition of Na plus K, and suggested that normal astrocytes lacked Na, K, Mg-ATPase activity as well as a Na pump. Those studies indicate that the ATPase activity is primarily associated with the neuronal rather than the glial element of the central nervous system.

The localization of ATPase activity in squid nerve fiber was histochemically determined by Sabatini et al. (143), who found that the activity was present mainly in axolemma toward its axoplasmic side and at the channels crossing the layer of Schwann cell membrane. Being specific to ATP, the enzyme did not release inorganic phosphate from ADP, AMP, or GTP, and, expecially in axolemma, was inhibited by G-strophanthin and K-strophanthoside, suggesting that the axolemma contained more Na, K, Mg-ATPase. An intense reaction for Mg-ATPase activity was shown histochemically in the perikaryon of the Purkinje cell of the cerebellum, and in numerous nerve fibers localized in the molecular layer surrounding the body of the Purkinje cell (144). Judging by their topography the fibers were basket cell fibers, and some of them seemed to be parallel fibers or the processes of stellate cells. The reaction in the fibers may well reflect the activity of Na, K, Mg-ATPase as well, since distinguishing Na, K, Mg-ATPase activity distinctly from Mg-ATPase activity is hard under some experimental conditions (145).

3. Nerve Endings.

It is of great interest that nerve ending fraction showed the highest specific activity of Na, K, Mg-ATPase among the fractions prepared from cerebral hemisphere of rat (128). The fact was substantiated by the report of Hosie (146) that a fraction consisting mainly of the external membranes of nerve endings of guinea pig brain was rich in Na, K, Mg-ATPase and that the specific activity of the fraction was higher than that of the conventionally prepared microsomal fraction. Kurokawa et al. (147) showed that approximately 20% of the ATPase activity originally present in the cerebral gray matter suspension was recovered in the fraction consisting principally of large nerve ending particles, concluding that the limiting membrane of the particle is one of the feasible sites of localization of the ATPase in cerebral tissues. These findings were confirmed by a histochemical technique (148). The presence of the ATPase naturally predicts the existence of active Na transport across the external membrane of the nerve endings. Indeed, in the experiments with [22]Na and various concentrations of K^+ and ouabain, Abdel-Latif (149) indicated that active Na-K transport took place in the nerve ending particles with intact synaptosomal membranes.

B. Acetylcholine Esterase.

Acetylcholine was reported to show interesting effects on Na, K, Mg-ATPase of microsomes of rat brain (150). Only in the presence of Na, K, and Mg, ATP and acetylcholine were found to be competitive inhibitors to acetylcholine esterase and ATPase, respectively, and the ouabain inhibition of acetylcholine esterase was maximum at the Na:K ratio which was optimal for the ATPase. The association of butyrylcholine esterase activity with Na transport was investigaged with the supraorbital salt-secreting gland of duck by Fourman (151), who noted that esterase activity was present in the epithelial cells of the gland only after stimulation by the in vivo administration of NaCl. The correlation of the ATPase to cholinesterases is of great interest and still remains to be understood.

C. Blood-brain Barrier

The administration of ATP into the cerebrospinal canal of dog decreased the glucose and glutathione content of the cerebrospinal fluid and increased Pi (152). It was suggested that the ATPase located in the cell membrane of the inner surface of the cerebrospinal canal was important in regulating the transport of substances across the barrier. The importance of the ATPase located in brain capillary in the regulation of the hemato-encephalic barrier was also suggested by Joo (153), who studied histochemically the inhibition of the enzyme after injection of nickel chloride in rats. The relation of Na, K, Mg-ATPase to blood-brain barrier, however, is yet to be clarified.

D. ATP Level and Nervous Function

Dependence of the function of the nervous system on ATP is shown by several studies. In chicken vagus nerves kept under nitrogen the ATP concentration and the compound action potential decreased rapidly; this occurred more rapidly when ATP regeneration via glycolysis was also inhibited (154). The spike height, conduction rate, and ATP content decreased proportionally with the amount of electrical stimulation. The ATP cost of nerve activity was calculated to be 5.4×10^{-10} moles/impulse/g wet weight. A similar experiment was also carried out with rabbit vagus nerve (155), and the ATP level was reported to decrease to 30% of the original resting

level, and the ADP level to increase simultaneously to 150% after the nerve
had been electrically stimulated for 25 sec (156).

Extending the same idea to the whole rat brain in vivo, Samson (157)
found that the rate of cerebral ATP utilization was slowed as cerebral
activity was reduced by lowering body temperature or by drugs. The
survival time of rats corresponded to the same concentration level of ATP
regardless of the extent of the inhibition of ATP regeneration. Assuming
that the survival time depended on the cerebral physiological function which
would be sustained by the normal ion gradient, he argued that ATP was a
main energy source to maintain the normal ion gradient across the neuronal
cell membrane.

He further argued that very small extracellular space of brain and
the enormous ramifications of the minute processes would mean that the
movement of small quantities of Na$^+$ and K$^+$ would produce sizeable
changes in ion gradients and that consequently the continuous operation
of cation pump was needed, though the size of "extracellular" space
where small ions move around relatively freely is still controversial
(142, 158). Though Na, K, Mg-ATPase, in general, plays a central role in
providing the internal environment of the cell with a constant physiological
condition by acting as an active transport mechanism, the ATPase system
serves a specific need of the nervous system, since there is little doubt
that the nerve conducting function is, under physiological conditions, fully
dependent on the electrochemical gradient sustained by the transport mechanism across the neuronal membrane (157). In this connection, it is worth
mentioning a report by Matsuda et al. (159) on the stimulatory effects of
ATP and some ions on the release of acetylicholine from synaptic vesicles
prepared from rat brain.

E. Development

Studies on developing rat brain demonstrated that Na, K, Mg-ATPase
activity enhanced during neonatal maturation along with dendritic formation
and an increase in electrical activity (160-162). The maturation from birth
to early adulthood was associated with a development of decreasing relationship of volume to active surface to which Na, K, Mg-ATPase is intimately
connected (157). Abdel-Latif et al. (163) reported that the ATPase activity
augmented abruptly on the 21st day of gestation, when the EEG was first
detectable, and reached the adult level by the 12th day post partum. Just
prior to this period, the synaptic components showed maturation and the
EEG became more rhythmic and regular. In the cerebral cortex of kitten
as well, a 12-fold increase in Na, K, Mg-ATPase activity was noted between
1 to 6 weeks, which was the period of rapid rise in spontaneous electrical
activity in cortical neurons (164). Since as early as age ten days

Na, K, Mg-ATPase could be increased above resting levels by electrical
stimulation of the cortex, the advent of electrical activity in immature
cortex seems to result in induction of the enzyme (164).

F. Effects of Drugs

That the normal function of the nervous system is maintained by normal
operation of the Na, K, Mg-ATPase system is supported also by pharmacological experiments. Fatty acids (4-12 carbons) inhibited rat brain N, K,
Mg-ATPase in vitro at concentrations similar to the intravenous levels
causing narcosis in vivo. This may be interpreted to show that the sleep-
like state seen after fatty acid injection was a result of inhibition of the
ATPase of the brain, i.e., interfering with membrane repolarization (165).
Gottlieb and Savran (166) reached a similar conclusion from their observation on nitrous oxide narcosis. That metrazol, a convulsant, increased the
activity of both Na, K, Mg-ATPase and K, Mg-phosphatase of rat brain
either in vivo or in vitro suggested an involvement of the enzyme system in
affected cerebral function (167).

The following drugs strongly altering nervous function were reported
to be all inhibitory in vitro at low concentrations: chloroform, chlorothi-
azide, chlorpromazine, ethacrynic acid, ethanol, ether, imipramine, procaine amide, proparanolol, quinidine (168-172). Strychnine was found to
activate both in vivo and in vitro Na, K, Mg-ATPase of rat and goat brains
(173).

G. Effects of Other Factors

Such relation of abnormal cerebral function with the altered ATPase
activity was observed also in the case of cortical lesions induced by freez-
ing with an ethyl chloride spray (174). Maximal EEG abnormality and the
increase of Na, K, Mg-ATPase activity appeared simultaneously in the area
of the lesion, and later the identical changes appeared in the cortex con-
tralateral to the lesion. Based on these data, an abnormality in alkali ion
transport in convulsive states, and the fact that the anticonvulsant effect of
a drug depended on its ability to accelerate Na^+ extrusion from the cell [see
(174)], it was suggested that the activity increase in Na, K, Mg-ATPase
reflected an adaptive response to an increase in influx of Na^+ into neuronal
structure. There was a direct relation between the decrease of ATP level
in the rat brain and the severity of the burns on the surface of the rear legs
of the same animal, and the inhibition of nervous activity during burn ex-
haustion and distrubance of nervous regulation of the animal's function were
explained by this ATP decrease (175).

Bowler and Duncan (176) found that the temperature sensitivities of Mg-ATPases prepared from the brains of frog and rat were similar. The Na, K, Mg-ATPase from the frog was active even below 5°, at which temperature the frog normally lives, while the Na, K, Mg-ATPase of the rat was virtually inactive at 0°. This may indicate that Na, K, Mg-ATPase is not associated with it.

ACKNOWLEDGMENTS

The present work was supported partly by research grants NS-07686 and NS-06827 from the United States Public Health Service. The editorial assistance of Mrs. Suzanne O'Brien is gratefully acknowledged.

REFERENCES

1. J. C. Skou, Biochim. Biophys. Acta, 23, 394 (1957).
2. L. W. Hokin and M. R. Hokin, Ann. Rev. Biochem., 32, 553 (1963)
3. J. D. Judah and K. Ahmed, Biol. Rev., Cambridge Phil. Soc. 39, 160 (1964).
4. I. M. Glynn, Sci. Basis Mec., Ann. Rev., 217 (1966).
5. R. W. Albers, Ann. Rev. Biochem., 36, 727 (1967).
6. P. C. Caldwell, Physiol, Rev., 48, 1 (1968).
7. A. Rothstein, Ann. Rev. Physiol., 30, 15 (1968).
8. P. F. Baker, A. L. Hodgkin, and T. I. Shaw, Nature, 190, 885 (1961).
9. I. Tasaki and M. Shimamura, Proc. Nat. Acad. Sci. U.S., 50, 619 (1962)
10. B. Katz, Nerve, Muscle, and Synapse, McGraw-Hill, New York, 1966.
11. G. N. Ling, A Physical Theory for the Living State, Blaisdell, New York, 1962.
12. C. Terner, L. V. Eggleston, and H. A. Krebs, Biochem. J., 47, 139 (1949).
13. A. L. Hodgkin and R. D. Keynes, J. Physiol. (London), 128, 28 (1955).
14. A. M. Shanes and M. D. Berman, J. Gen. Physiol., 39, 279 (1955).
15. H. A. Pappius and K. A. C. Elliott, Can. J. Biochem. Physiol., 34, 1053 (1956).
16. F. J. Brinley, Jr. and L. J. Mullins, J. Gen. Physiol., 50, 2303 (1967).
17. P. C. Caldwell, J. Physiol. (London), 152, 545 (1960).
18. P. C. Caldwell, A. L. Hodgkin, R. D. Keynes, and T. I. Shaw, J. Physiol. (London), 152, 561 (1960).
19. F. J. Brinley, Jr. and L. J. Mullins, J. Gen. Physiol., 52, 181 (1968).

20. L. J. Mullins and F. J. Brinley, J. Gen. Physiol., 50, 2333 (1967).
21. F. J. Brinley, Jr., J. Gen. Physiol., 51, 149S (1968).
22. T. Nakao, Y. Tashima, K. Nagano, and M. Nakao, Biochem. Biophys. Res. Commun., 19, 755 (1965).
23. P. F. Baker, R. F. Foster, D. S. Gilbert, and T. I. Shaw, Biochim. Biophys. Acta, 163, 560 (1968).
24. P. F. Baker and J. Manil, Biochim., Biophys, Acta, 150, 328 (1968).
25. R. A. Sjodin and L. A. Beauge, J. Gen. Physiol., 51, 152S (1968).
26. H. P. Rang and J. M. Ritchie, J. Physiol. (London), 196, 163 (1968).
27. H. P. Rang and J. M. Ritchie, J. Physiol. (London), 196, 183 (1968).
28. H. H. Wespi, Pfluegers Arch. Ges. Physiol., 306, 262 (1969).
29. R. Whittam, Biochem. J., 84, 110 (1962).
30. P. F. Baker, Biochim, Biophys. Acta, 75, 287 (1963).
31. E. Giacobini, S. Hovmark, and Z. Kometiani, Acta Physiol. Scand., 71, 391 (1967).
32. P. C. Caldwell and R. D. Keynes, J. Physiol. (London), 148, 8P (1959).
33. E. T. Dunham and I. M. Glynn, J. Physiol. (London), 156, 274 (1961).
34. I. M. Glynn, Pharmacol, Rev., 16, 381 (1964).
35. S. L. Bonting and M. R. Canady, Am. J. Physiol., 207, 1005 (1964).
36. F. C. Herrera, Am. J. Physiol., 210, 980 (1966).
37. J. S. Charnock and R. L. Post, Nature, 199, 910 (1963).
38. J. C. Skou, Physiol. Rev., 45, 596 (1965).
39. G. D. V. van Rossum, Biochem. J., 84, 35P (1962).
40. R. Whittam, K. P. Wheeter, and A. Blake, Nature, 203, 720 (1964).
41. A. Blake, D. P. Jeader, and R. Whittam, J. Physiol. (London), 193. 467, (1967),
42. P. J. Garrahan and I. M. Glynn, J. Physiol. (London), 192, 237 (1967).
43. T. Valcana and P. S. Timira, J. Neurochem., 16, 935 (1969).
44. B. B. Gallagher and G. H. Glaser, J. Neurochem., 15, 525 (1968).
45. R. W. Albers, S. Fahn, and G. J. Koval, Proc. Nat. Acad. Sci. U.S., 50, 474 (1963).
46. K. Ahmed and J. D. Judah, Biochim. Biophys, Acta, 104, 112 (1965).
47. L. E. Hokin, P. S. Sastry, P. R. Galsworthy, and A. Yoda, Proc. Nat. Acad. Sci. U.S., 54, 117 (1965).
48. K. Nagano, T. Kanazawa, N. Mizuno, Y. Tashima, T. Nakao, and M. Nakao, Biochem., Biophys, Res., Commun,, 19, 759 (1965).
49. R. L. Post, A. K. Sen, and A. S. Rosenthal, J. Biol. Chem., 240 1437 (1965).
50. A. K. Sen and R. L. Post, Abstr. Am. Biophys. Soc. 10th Ann. Meeting, Boston, 1966, p. 152 [Biophys. J., 6 (1966)].
51. H. Bader, A. K. Sen, and R. L. Post, Biochim. Biophys. Acta, 118, 106 (1966).

52. L. J. Opit and J. S. Charnock, Nature, 208, 471 (1961).
53. J. C. Skou, Progr. Biophys. Molecular Biol. 14, 113-163 (1964).
54. R. E. Davies and R. D. Keynes, in Membrane Transport and Metabolism (A. Kleinzeller and A. Kotyk, eds.), Academic, New York, 1961, pp. 336-340.
55. H. McIlwain, in Ciba Foundation Symposium, Enzymes and Drug Action (A. V. S. de Reuck and J. L. Mongar, eds.), Churchill, London, 1962, p. 170.
56. R. L. Post, A. K. Sen, and A. S. Rosenthal, J. Biol. Chem., 240 1437 (1965).
57. G. R. Kepner and R. I. Macey, Biochim. Biophys. Acta, 163, 188 (1968).
58. N. Mizuno, K. Nagano, T. Nakao, Y. Tashima, M. Fujita, and M. Nakao, Biochim. Biophys, Acta, 168, 311 (1969).
59. M. Nakao, K. Nagano, T. Nakao, N. Mizuno, Y. Tashima, M. Fujita, H. Maeda, and H. Matsudaira, Biochem. Biophys. Res. Commun, 29, 588 (1967).
60. S. Uesugi, A. Kahlenberg, F. Medzihradsky, and L. E. Hokin, Arch. Biochem. Biophys., 130, 156 (1969).
61. J. F. Danielli and H. Davson, J. Cellular Comp. Physiol., 5, 495 (1935).
62. H. Davson and J. F. Danielli, The Permeability of Natural Membranes, 2nd ed., Cambridge Univ. London, 1952.
63. J. D. Robertson, in Cellular Membranes in Development (M. Locke, ed.) Academic, New York, 1964, pp. 1-81.
64. H. Fernández-Morán, Circulation, 26, 1039 (1962)
65. F. S. Sjöstrand, J. Ultrastruct. Res., 9, 340 (1963).
66. F. S. Sjöstrand, Nature, 199, 1262 (1963).
67. F. S. Sjöstrand and L.-G. Elfvin, J. Ultrastruct. Res., 10, 263 (1964).
68. S. E. G. Nilsson, Nature, 202, 509 (1964).
69. E. L. Benedetti and P. Emmeiot, J. Cell Biol., 26, 299 (1965).
70. W. Stoecekenjus and S. C. Mohr, Lab. Invest., 14, 1198 (1965).
71. D. Branton, Proc. Nat. Acad. Sci. U. S., 55, 1048 (1966).
72. S. Fleischer, B. Fleischer, and W. Stoeckenius, J. Cell Biol, 32, 193 (1967).
73. J. P. Revel and M. J. Karnovsky, J. Cell Biol., 33, C7 (1967).
74. D. E. Green, D. W. Allmann, E. Bachmann, K. Kopaczyk, E. F. Korman, S. Liption, H. Baum, D. H. Maclennan, D. G. McConnell, J. F. Perdue, J. S. Rieskye, and A. Tzagoloff, Arch. Biochem. Biophys., 119, 312 (1967).
75. J. K. Blasie, M. M. Dewey, A. E. Blanrock, and C. R. Worthington, J. Mol. Biol., 14, 143 (1965).
76. A. H. Maddy and B. R. Malcolm, Science, 150, 1616 (1965).
77. J. B. Finean, R. Coleman, W. G. Green, and A. R. Limbrick, J. Cell Sci., 1, 287 (1966).

78. J. Lenard and S. J. Singer, Proc. Nat. Acad. Sci. U. S., 56, 1828 (1966).

79. V. Luzzati, F. Reiss-Husson, E. Rivas, and T. Gulik-Kryzywicki, Ann. N. Y. Acad. Sci., 137, 409 (1966).

80. A. H. Maddy and B. R. Malcolm, Science, 153, 212 (1966).

81. D. F. H. Wallach and P. H. Zahler, Proc. Nat. Acad. Sci. U. S., 56, 1552 (1966),

82. D. E. Green and S. Fleischer, Biochim. Biophys, Acta, 70, 554 (1963).

83. J. A. Lucy, J. Theoret, Biol, 7, 360 (1964).

84. J. L. Kavanau, Structure and Function in Biological Membranes, Holden-Day, San Francisco, 1965.

85. J. S. O'Brien, J. Theoret, Biol., 15, 307 (1967).

86. B. Ke, Arch. Biochem. Biophys., 112, 554 (1965).

87. B. Ke, Nature, 208, 573 (1965).

88. J. M. Steim, 153rd Meeting, American Chemical Society, Miami Beach, Florida, April 1967.

89. J. M. Steim and S. Fleischer, Proc. Nat. Acad. Sci. U. S., 58. 1292 (1967).

90. D. W. Urry, M. Mednieks and E. Benjnarowicz, Proc. Nat. Acad. Sci. U. S. 57, 1043 (1967).

91. D. F. H. Wallach and A. Gordon, in Protides of the Biological Fluids (H. Peeters, ed.), Vol. 15, Elsevier, Amsterdam, 1967, p. 47.

92. D. F. H. Wallach and P. H. Zahler, Biochim. Biophys. Acta, 150, 186 (1968).

93. D. Chapman, V. B. Kamat, J. de Gier, and S. A. Penkett, Nature, 213, 74 (1967).

94. D. Chapman, V. B. Kamat, J. de Gier, and S. A. Penkett, J. Mol. Biol. 31, 101 (1968).

95. J. C. Skou, in Membrane Transport and Metabolism (A. Kleinzeller and A. Kotyk, eds.), Academic, New York, 1961, p. 228.

96. M. Tachibana, J. Biochem. (Tokyo), 53, 260 (1963).

97. T. Narahashi and J. M. Tobias, Am. J. Physiol., 207, 1441 (1964).

98. T. Ohonishi and H. Kawamura, J. Biochem. (Tokyo), 56, 377 (1964).

99. I. Zamudio, M. Cellino, and M. Canessa-Fischer, Arch. Biochem. Biophys., 129, 336 (1969).

100. H. J. Schatzmann, Nature, 196, 677 (1962).

101. J. B. Finean and A. Martonosi, Biochim. Biophys. Acta, 98, 547 (1965).

102. E. L. Benedetti and P. Emmelot, J. Microsc., 5, 645 (1966).

103. R. J. Lesseps, J. Cell Biol., 34, 173 (1967).

104. J. Lenard and S. J. Singer, Science, 159, 738 (1968).

105. L. Lerman, A. Watanabe, and I. Tasaki, Neuroscience Res., 2, 71 (1969).

106. A. D. Brown, J. Mol. Biol., 12, 491 (1965).

107. C. W. F. McClare, Nature, 216, 766 (1967).
108. S. H. Richardson, Biochim. Biophys, Acta, 150, 572 (1968).
109. K. P. Wheeler and R. Whittam, Biochem. J., 93, 349 (1964).
110. J. J. Kabara and D. Konich, 41st Fall Meeting, American Oil Chemists Society, Chicago, Ill., October 1967.
111. R. Tanaka and L. G. Abood, Arch. Biochem. Biophys., 108, 47 (1964).
112. R. Tanaka and K. P. Strickland, Arch. Biochem. Biophys., 111, 583 (1965).
113. R. Tanaka and L. G. Abood, J. Neurochem., 10, 571 (1963).
114. D. M. Small and M. Bourgés, Molecular Crystals, 1, 173 (1966).
115. R. Tanaka, J. Neurochem., 16, 1301 (1969).
116. R. Tanaka and T. Mitsumata, J. Neurochem., 16, 1163 (1969).
117. R. Tanaka and T. Sakamoto, Biochim. Biophys. Acta, 193, 384 (1969).
118. M. Dixon and C. Webb, Enzymes, 2nd ed., Academic, New York, 1964, Chap. 6.
119. E. E. Jacobs, M. Jacobs, D. R. Sanadi, and L. B. Bradley, J. Biol. Chem., 223, 147 (1956).
120. A. L. Fluharty and R. Sanadi, J. Biol. Chem., 236, 2772 (1961).
121. R. Tanaka, T. Sakamoto, and Y. Sakamoto, J. Membrane Biol. 4, 42 (1971). R. Tanaka and T. Sakamoto, unpublished data.
122. J. D. Judah, K. Ahmed, and A. E. M. McLean, Biochim. Biophys. Acta, 65, 472 (1962).
123. K. Ahmed and J. D. Judah, Biochim. Biophys. Acta, 93, 603 (1964).
124. M. Fujita, T. Nakao, Y. Tashima, N. Mizuno, K. Nagano, and M. Nakao, Biochim. Biophys. Acta, 117, 42 (1966).
125. Y. Hashiuchi, Nara Igaku Zasshi, 18, 469 (1967).
126. B. Formby and C. Joergen, Z. Physiol, Chem., 349, 909 (1968).
127. A. F. Rega, P. J. Garrahan, and M. I. Pouchan, Biochim. Biophys. Acta, 150, 742 (1968).
128. R. W. Albers, G. Rodriguez de Lores Arnaiz, and E. de Robertis, Proc. Nat. Acad. Sci. U. S., 53, 557 (1965).
129. R. W. Albers and G. J. Koval, J. Biol. Chem., 241, 1896 (1966).
130. H. Barder and A. K. Sen, Biochim. Biophys. Acta, 118, 116 (1966).
131. F. Nagai, F. Izumi, and H. Yoshida, J. Biochem., 59, 295 (1966).
132. H. Yoshida, F. Izumi, and K. Nagai, Biochim. Biophys. Acta, 120, 183 (1966).
133. C. E. Inturrisi, Biochim. Biophys, Acta, 173, 567 (1969).
134. N. M. Mirsalikhova, S. A. Temirov, and B. A. Tashmukhamedov, Uzbeksk. Biol. Zh., 12, 63 (1968) (in Russian); through CA, 70, 45188d (1969).
135. S. L. Bonting and L. L. Caravaggio, Arch. Biochem, Biophys., 101, 37 (1963).
136. S. L. Bonting, K. A. Simon, and N. M. Hawkins, Arch. Biochem. Biophys., 95, 416 (1961).

137. S. Fahn and L. J. Côté, J. Neurochem., 15, 433 (1968).
138. T. Harmony, R. Urba-Holmgren, and C. M. Urbay, Brain Res., 5, 109 (1967).
139. V. T. Marchesi and G. E. Palade, J. Cell Biol., 35, 385 (1967).
140. J. Cummins and H. Hydén, Biochim. Biophys. Acta, 60, 271 (1962).
141. E. Lewin and H. H. Hess, J. Neurochem., 11, 471 (1964).
142. H. H. Hess, G. Schneider, M. Warnock, and A. Pope, Federation Proc., 22, 333 (1963).
143. M. T. Sabatini, R. Dipolo, and R. Villegas, J. Cell Biol., 38, 176 (1968).
144. D. Onicescu and I. Cuida, J. Neurochem., 16, 467 (1969).
145. R. Tanaka and L. G. Abood, Arch. Biochem. Biophys., 105, 554 (1964).
146. R. J. A. Hosie, Biochem. J., 96, 404 (1965).
147. M. Kurokawa, T. Sakamoto, and M. Kato, Biochem. J., 97, 833 (1965).
148. V. F. Mashanskii and A. S. Mirkin, Dokl. Akad. Nauk. SSSR, 184, 1423 (1969) (in Russian); through CA, 70, 111712y (1969).
149. A. A. Abdel-Latif, J. Neurochem., 15, 721 (1968).
150. Z. P. Kometiani and A. A. Kalandarishvili, Biofizika, 14, 213 (1969) (in Russian); through CA, 70, 111587m (1969).
151. J. Fourman, J. Anat., 104, 233 (1969).
152. A. S. Oganesyan, Zh. S. Gevorkyan, Arm. Khim. Zh., 21, 75 (1968) (in Russian); through CA, 69, 34094x (1968).
153. F. Joo, Nature, 219, 1378 (1968).
154. N. A. Dahl, F. E. Samson, Jr., and W. M. Balfour, Am. J. Physiol., 206, 818 (1964).
155. P. Greengard and R. W. Straub, J. Physiol. (London) 148, 353 (1959).
156. P. Montant and M. Chmouliovsky, Experientia, 24, 782 (1968).
157. F. E. Samson, Jr., Progr. Brain Res., 16, 216 (1965).
158. J. Dobbing, Physiol. Rev., 41, 130 (1961).
159. T. Matsuda, F. Hata, and H. Yoshida, Biochim. Biophys, Acta, 150, 739 (1968).
160. L. J. Côté, Life Sci., 3, 899 (1964).
161. F. E. Samson, Jr., H. Dick, and W. M. Balfour, Life Sci., 3, 511 (1964).
162. F. E. Samson, Jr. and D. J. Quinn, J. Neurochem., 14, 421 (1967).
163. A. A. Abdel-Latif, J. Brody, and H. Raman, J. Neurochem., 14 1133 (1967).
164. P. R. Huttenlocher and M. D. Rawson, Exp. Neurol., 22, 118 (1968).
165. D. R. Dahl, J. Neurochem., 15, 815 (1968).
166. S. F. Gottlieb and S. V. Savran, The Physiologist, 9, 192 (1966).

167. E. de Robertis, M. Alberici, and G. Rodriguez de Lores Arnaiz, Brain Res., 12, 461 (1969).
168. R. F. Squires, Biochem. Biophys. Res. Commun., 19, 27 (1965).
169. Y. Israel and I. Salazar, Arch. Biochem. Biophys., 122, 310 (1967).
170. U. Tarve and M. Brechtlova, J. Neurochem., 14, 283 (1967).
171. J. D. Robinson, J. Lowinger, and B. Bettinger, Biochem. Pharmacol., 17, 1113 (1968).
172. S. Y. Song and J. Scheuer, Pharmacology, 1, 209 (1969).
173. B. K. Pal and J. J. Ghosh, J. Neurochem., 15, 1243 (1968).
174. E. Lewin and A. McCrimmon, Arch. Neurol., 16, 321 (1967).
175. V. V. Davydov, Patol. Fiziol. i. Eksperim. Terapiya, 12, 39 (1968) (in Russian); through CA, 70, 18231f (1969).
176. K. Bowler and C. J. Duncan, Comp. Biochem. Physiol., 24, 223 (1968).
177. R. Tanaka and H. Morita, unpublished data.
178. L. Weil, S. James, and A. R. Buchert, Arch. Biochem. Biophys., 46, 266 (1953).

NOTE ADDED IN PROOF

Addendum to Sec. II. C. General Properties of ATPase

5. Inhibitors

As described on p. 55, it has been concluded that sulfhydryl groups are involved in the activity of the ATPase but that the groups are not phosphorylated during the hydrolysis reaction of ATP. Since histidine, besides serine, is a well known amino acid residue that is chemically active and intimately associated with some enzyme reactions, we tested the involvement of the amino acid residue in the ATPase activity using a solubilized preparation from bovine cerebral cortex (177).

(a) When the ATPase preparation was mildly oxidized by photoirradiation in the presence of methylene blue under oxygen or air the activity was lost according to the time of the irradiation. The loss of the activity was not prevented by the addition of either ATP or crude licithin preparation. Inasmuch as the amino acid residues oxidized under these conditions were proved to be histidine and tryptophan, and neither denaturation nor hydrolysis of peptide bond was detected (178), the loss of the ATPase activity seemed to be due to oxidation of those amino acid residues. (b) A treatment of the ATPase preparation with p-diazobenzene sulfonate reduced the enzyme activity. It has been suggested that the diazo compound reacts with imidazole, phenol, and possibly indole, but the loss of the ATPase activity was protected in part by the addition of either histidine or cysteine to a similar extent. Thus, the results indicated that either histidine or cysteine or both are involved in the activity. (c) Cu^{2+} is a well known ion to bind tightly to cysteine and N-terminal histidine residues. Because the ion strongly inhibited the ATPase and the activity was recovered to a similar extent by the addition of histidine or cysteine, those two amino acid residues are again likely to be associated with the enzyme activity. (d) The hydroxyl group of the serine residue did not appear to be related to the active center of the enzyme because no inhibition was shown even in the presence of 10^{-3} M diisopropylfluorophosphate, which reacts with the serine residue of protein.

These experimental results added strong support to the association of cysteine residue with the activity of the ATPase, and the involvement of histidine residue was also well supported.

Chapter 3

NERVE EXCITATION AND "PHASE TRANSITION" IN MEMBRANE MACROMOLECULES

Ichiji Tasaki

Laboratory of Neurobiology
National Institute of Mental Health
Bethesda, Maryland 20014

I. INTRODUCTION

The action potential is an electric manifestation of a series of physico-chemical events occurring in a thin lipoprotein layer, the axolemma, at the surface of an excitable cell. There is little doubt that the macromolecules which constitute the axolemma undergo drastic reversible conformational changes during the action potential. However, because of the extremely small volume occupied by the axolemma, it is not easy to follow these changes by the use of various instruments commonly employed in biochemical investigation of macromolecules. The high sensitivity and the high time resolution required for instruments used for study of excitation processes are undoubtedly the direct consequence of the extreme thinness of the axolemma.

Another factor which makes biochemical and physicochemical studies of the axolemma very difficult is the lability of the membrane macromolecules. It is well known that excitability of a nerve can be irreversibly eliminated by various physical and chemical agents. Slight compression or stretch, immersion in hypo- or hypertonic media or in isotonic but unbalanced salt solutions, application of weak acid or alkali, etc., can permanently destroy the normal architecture of the membrane macromolecules. Furthermore, salts of divalent cations, which are essential in the external medium to the process of nerve excitation, are known to bring about an immediate and irreversible loss of excitability when applied intracellularly at low concentrations.

In connection with the lability of the axonal membrane, it should be pointed out that the presence of excitability is related, in a sense, to the labile and unstable character of the membrane. Under ordinary experimental conditions, the change in the transmembrane potential required to evoke an action potential is smaller than kT/e (i.e., 25.2 mV at 20° C), where k is the Boltzmann constant, T the absolute temperature and e the electronic charge. Since the average kinetic (translational or rotational) energy of mobile ions or charged groups of the membrane macromolecules is of the order of kT, it is evident that the energy delivered by the applied stimulus is by itself insufficient to totally alter the pattern of random thermal motion of the charged groups of membrane macromolecules or Brownian motion of interdiffusing ions. In fact, the energy derived directly from the stimulating pulse is insufficient to break even a weak chemical bond, such as a hydrogen bond. Therefore, it is difficult to ascribe structural changes in the membrane to a direct effect of the applied voltage, per se. The reason why such a weak electric pulse brings about a drastic structural change in the membrane may be readily understood if the resting state of the nerve membrane is assumed to be close to an unstable state of the system where an autocatalytic, or cooperative, process within the membrane can be initiated.

In this article, an attempt is made to review the results of recent experiments designed to elucidate physicochemical properties of the nerve membrane. The major portion of the findings described was obtained by the use of intracellular perfusion technique in conjunction with the standard electrophysiological technique developed for investigating these labile systems. These results are an extension of the work done by Cole (1), Hodgkin (2), and by others with improved techniques. Therefore, it seems appropriate to begin this review with a discussion of the role of the various ions in the process of excitation.

In the field of physical chemistry, every investigator tries to simplify the experimental conditions and to reduce the number of variables affecting

the results of his measurements. Excitable membranes are located between the external fluid medium and the protoplasm in the interior of the
cell. In small cells in which the cell interior is inaccessible to direct
experimental manipulation, the number of variables affecting the results
is very large and the types of physico-chemically meaningful measurements
that can be carried out are quite limited. Therefore, it is difficult, if not
impossible, to build a physicochemical theory of nerve excitation based on
the observations using these small cells. For this reason, the entire
argument developed in this article is founded on the experimental data obtained from squid giant axons, where the internal as well as external milieu
may be controlled experimentally.

By pursuing this process of simplification of the experimental conditions further, it has become evident that analysis of a membrane system
containing only two different species of cations is far easier and less amenable to misinterpretation of the results than that of a system with three or
more cation species. For this reason, a great emphasis is placed in later
sections on the results of analysis of action potentials observed under "bi-
ionic" conditions. Naturalistic investigators may call these conditions
"abnormal." However, it should be noted that the objective of our investigation lies in elucidation of the excitation mechanism in the nerve membrane
on molecular basis, and not in mere description of potentials and currents
measured under "normal" conditions. As long as our measurements are
carried out on axon membranes showing no sign of irreversible alteration,
changes in the membrane macromolecules should be limited to their secondary and tertiary structures. Hence the results obtained may be safely
utilized to interpret the behavior of the membrane macromolecules under
normal, i.e., multi-ionic conditions.

II. Na-SUBSTITUTION IN THE EXTERNAL MEDIUM

Natural sea water contains approximately 110 meq/liter of divalent
cations (Ca and Mg) and about 500 meq/liter of univalent cations, mainly
sodium. Reduction of the sodium ion concentration by replacing sodium
with polyatomic univalent cation surrounded by hydrophobic groups decreases the action potential amplitude in accordance with the Nernst equation

$$E_a = \frac{kT}{e} \ln [Na]_e + \text{constant} \tag{1}$$

where E_a is the action potential amplitude of an axon immersed in external
medium with an external sodium concentration $[Na]_e$ and kT/e has the
usual thermodynamic meaning. This relationship is valid only in the range

where the sodium concentration is larger than the divalent cation concentration of the medium. This rule is an extension of the finding by Hodgkin and Katz (3) and is applicable also to squid giant axons perfused with various internal solutions (4).

It is well known that sodium ion in aqueous media is highly hydrated [see Frank and Wen (5), Millero (6)]. Electrostatic attraction between the charge of the ion and the dipole of water molecules creates layers of water molecules surrounding the instrinsic part of the ion. Lithium ions are very similar to sodium ions in this respect and are known to be effective in substituting sodium ions in the external medium [Hodgkin and Katz (3)].

Among polyatomic ions, there are a number of univalent cations which have hydrophilic side groups, such as NH_2 or OH. Lorente de No, Vidal, and Laramendi (7) found that many of these organic cations are actually favorable sodium substitutes in frog nerve fibers. More recently, it was found that such univalent cations as guanidine, hydrazine, hydroxylamine, aminoguanidine, etc., are excellent sodium substitutes in squid giant axons under intracellular perfusion (8). In an external medium containing the salt of any one of these univalent cations and a calcium salt, the action potential amplitude was found to vary in accordance with Eq. (1) in which $[Na]_e$ was replaced with the concentration of the polyatomic cation.

Figure 1 illustrates the effect of substitution of NaCl in the external medium (containing 200 mM $CaCl_2$) first with choline chloride and subsequently with hydrazine chloride. The second and third records in this figure show that there is some sign of excitability in an internally perfused axon immersed in a mixture of choline chloride and $CaCl_2$. The sodium substituting capability of a polyatomic cation decreases with the number of hydrophobic groups attached to the nitrogen atom. There are many ions which are intermediate between hydrazine and choline in their sodium substituting capability. The details of the experimental results showing the relationship between the chemical structure and the sodium substituting capability may be found elsewhere (9).

The difference in behavior between univalent cations with hydrophobic side groups and those with hydrophilic side groups may be attributed to a special property of the axon membrane. It has been suggested that the outer layer of the membrane may undergo, in response to stimulation, a transition from a dense, more-or-less hydrophobic state to a new, less dense state with an increased water content (4). Such a transition would increase the selectivity of the outer layer of the axon membrane for cations with hydrophilic side groups, but not for hydrophobic cations. Externally applied cation with high selectivity are expected to strongly influence the membrane potential (see later sections).

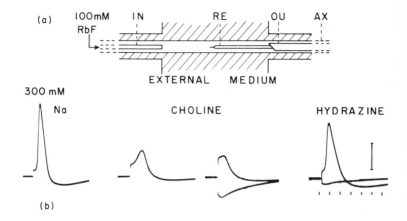

Fig. 1. (a): Experimental arrangement (not to scale) used for demonstration of excitability of squid giant axon immersed in sodium free medium and internally perfused with 100 mM RbF. AX, giant axon; RE, recording electrode (wire enclosed in glass capillary); OU, outlet cannula, IN; inlet cannula. (b): An example of the action potential records obtained with the above arrangement. The composition of the external fluid medium was 300 mM NaCl and 200 mM CaCl$_2$ (left record), 300 mM choline chloride and 200 mM CaCl$_2$ (center records) and 300 mM hydrazine chloride and 200 mM CaCl$_2$ (right record). The latter two media were sodium free. The bar represents 50 mV and the time markers are 1 msec apart. The perfusion zone was 12 mm in length, and the axon diameter was 620$_\mu$. (From Tasaki, Singer, and Watanabe, Proc. Natl. Acad. Sci. U.S. 54, 765, 1965.)

III. BI-IONIC ACTION POTENTIALS

Soon after the nonessentiality of sodium ion in excitation of the squid axon was established, another noteworthy experimental finding was reported, shedding new light on the mechanism of excitation. It was found that, under intracellular perfusion with various favorable salt solutions, squid axons maintain their ability to develop action potentials in an external medium containing a divalent cation salt as the sole electrolyte (10, 11). As the result of this and the subsequent studies, it is now established that the presence of univalent cations (such as Na or its substitutes) in the external medium is not required for production of action potentials (12, 13).

Figure 2 illustrates the action potentials obtained from an axon with a CaCl$_2$ solution externally and a CsF solution internally. (Glycerol was used to maintain the tonicity inside and outside the membrane.) In other experiments, salts of Sr or Ba were used in place of Ca; the ability of the

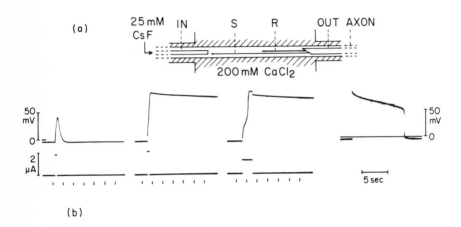

(b)

Fig. 2. (a): Schematic diagram showing the experimental arrangement used to obtain the oscillograph records shown in (b). The inlet cannula (IN), outlet cannula (OUT), stimulating wire electrode (S), and recording glass capillary electrode (R) are shown. The distance between the tips of the twó cannula was approximately 20 mm. The diameter of the axon was approximately 500μ. (b): Records of action potentials observed in the absence of univalent cation in the external fluid medium. In the first record (extreme left), the stimulating current pulse (indicated by the lower oscillograph trace) was subthreshold. The time markers are 50 msec apart. The last record was taken at a slower sweep speed. The zero level for the potential trace indicates the potential observed when the recording electrode (filled with 0.6 M KCl) was placed in the outside fluid medium. Temperature 21° C. (From Tasaki, Watanabe, and Singer. Proc. Natl. Acad. Sci. U.S., 56, 1116, 1966.)

axon to develop action potentials was maintained with these divalent cations externally. Attempts were made without success to obtain action potentials in axons immersed in a medium containing only univalent cation salt. In a medium containing the salt of sodium ion only, with or without EDTA for chelation of Ca, axons were found to lose their excitability in a relatively short period of time.

In this type of experiment, the internal univalent cation, cesium ion in the experiment illustrated in Fig. 2, can be replaced with sodium, lithium, choline, tetramethylammonium, tetraethylammonium, guanidine, etc., without suppressing excitability. As a rule, internal univalent cations with hydrophobic side groups (such as CH_3 -, C_2H_5 -, etc.) give rise to larger action potentials than hydrated cations (such as Na).

The chemical species of the anions in the medium can be varied to a great extent without losing excitability. Bromide, ethylsulfate and others, which do not interfere with the calcium ion, have been used externally, yielding essentially the same results. Anions with low lyoptropic numbers (fluoride, phosphate, glutamate, etc.) are favorable internally; those near the opposite extreme in Hoffmeister's series are known to show a strong tendency to destroy the normal structure of the axon membrane (14). There is good evidence that the membrane macromolecules in the physiological range of pH (7, 8) show properties of polyanions. The negative charge of the membrane macromolecules tends to expel small anions derived from the external and internal fluid media. From this and other studies, it is inferred that anions are not directly involved in the process of action potential production.

There are a number of similarities between an axon membrane and an inanimate cation exchange membrane. Changes in the membrane potential brought about by changing the external salt concentration have the same sign as those seen in cation exchange membrane (11, 15). Therefore, the potential difference across the axon membrane with two different cation species, one species on each side, is in some respects analogous to what physical chemists call "bi-ionic potential" (16, 17). For this reason, the action potentials involving only two cation species, divalent cation externally and univalent cation internally, are often designated as "bi-ionic action potentials."

IV. EXCITATION OF AXONS INTERNALLY PERFUSED WITH SODIUM SALT SOLUTION

In an axon immersed in a medium containing a divalent cation salt of a given concentration, the amplitude of the bi-ionic action potential varies with the concentration and the chemical species of the internal univalent cation. With a medium of 100 mM $CaCl_2$ externally, the action potential amplitude of an axon internally perfused with a 30 mM sodium-phosphate solution is about 60 mV (see Fig. 3). This value is smaller than that obtained with tetramethylammonium or tetraethylammonium internally; but, it is larger than the values obtained with guanidine or rubidium inside. This fact indicates that sodium ion does not occupy any special extreme position in the list of internal univalent cations tested. In other words, these observations do not support the view that the action potential is produced by an increase in membrane permeability specifically to sodium ion.

Electric properties of the squid axon membrane under the bi-ionic conditions have been examined by the voltage-clamp technique. With a dilute sodium phosphate solution internally and a $CaCl_2$ solution externally, the

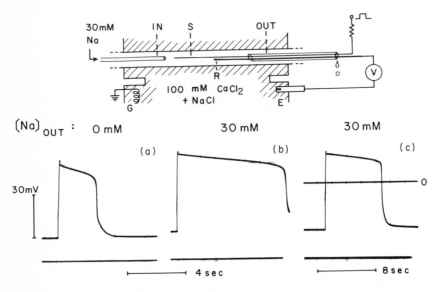

Fig. 3. Resting and action potentials recorded from an axon internally perfused with a 30 mM sodium salt solution. The external media
contained 100 mM $CaCl_2$ for all the records. The external sodium concentrations were varied as indicated. The internal anion was a 1:1 mixture
of fluoride and phosphate in this case. Stimuli used were 100 msec in
duration and approximately $1.5 \mu A/cm^2$ in intensity (indicated by the lower
trace). Note that record (c) was taken at a slower sweep speed. In record
(c), the potential recorded when the internal recording electrode was with-
drawn and placed in the external medium was superposed on the action
potential trace. Axon diameter: approximately 450μ. Temperature, 20°
C. (From Watanabe, Tasaki, and Lerman, Proc. Natl. Acad. Sci. U.S.,
58, 2246, 1967.)

relationship between the transmembrane potential, V, and the current, I,
was found to be qualitatively similar to those encountered in unperfused
axons.

Figures 4 and 5 shows an example of the experimental results obtained
by the use of the voltage-clamp technique applied to an axon under bi-ionic
conditions. Symbol V in these figures represents the level of transmembrane potential artificially maintained above (and below) the resting potential by an automatic control of the membrane current. A positive value of
V shows that the intracellular potential (relative to that of the external
medium) was suddenly raised above the resting level by the amount indicated. In electrophysiology, the term "depolarization" is used to denote
a rise in the intracellular potential. Changes in the transmembrane potential in the reverse direction are called "hyperpolarization."

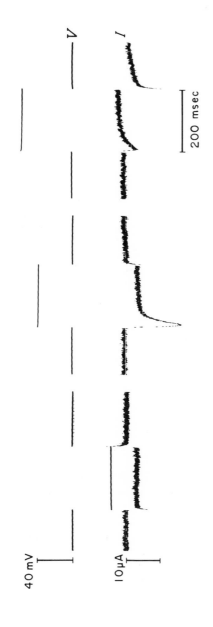

Fig. 4. Membrane potentials and currents recorded under voltage-clamp of an axon internally perfused with sodium phosphate and immersed in CaCl₂ solution. Record 1 is the current-time trace obtained when the voltage pulse used to clamp the axon membrane was hyperpolarizing. Record 2 shows the current-time trace of the membrane current when the clamping pulse was depolarizing and 43 mV in amplitude. In record 3, the current-time trace observed when the clamping voltage was 56 mV in amplitude and depolarizing in direction is presented. Axon diameter: approximately 450 μ. Temperature, 19° C. (From Tasaki, Lerman, and Watanabe, Am. J. Physiol., 216, 130, 1969.

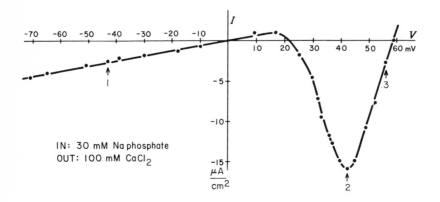

Fig. 5. Current-voltage relationship obtained from squid axon, internally perfused with sodium phosphate, in external media containing
$CaCl_2$ as the sole electrolyte species. The V-shaped curve represents
the relationships between the peak inward current (I) and the clamping
voltage (V). Three examples of the oscillograph record (1, 2, and 3) from
which this I-V relationship is constructed are given in Fig. 4. (From
Tasaki, Lerman, and Watanabe, Am. J. Physiol., 216, 130, 1969.)

When hyperpolarizing clamping pulses were used, the steady level of
the membrane current, I, was found to vary directly with the magnitude of
potential shift, V. In other words, the axon membrane obeys Ohm's law in
the negative range of V. In accordance with the definition of V, the straight
line representing the linear relationship between V and I passes through the
origin of the I-V diagram.

There are other points in the observed I-V relationship at which membrane current I is zero. When the clamping potential level is close to the
peak value of the bi-ionic action potential (about 60 mV), the membrane current vanishes following the so-called "capacitative surge." This finding
can easily be understood on the basis of the following consideration:

Under the ordinary, unclamped conditions, the threshold (critical)
potential level required for excitation is about 20 mV above the resting
potential. Since the peak level of the action potential (approximately 60
mV) is far higher than the threshold level, the axon membrane becomes
active (or excited) shortly after the membrane potential is raised to the
peak level of the action potential. The state of the membrane reached at a
peak of excitation is characterized by $I = O$ and $V = E_a$, where E_a is the
peak level of the action potential. Therefore, it is expected that the membrane current, I, vanishes when V is approximately 60 mV.

The resistance of the membrane in the excited state can be determined from the slope of the linear I-V relationship near the point where I = O and $V = E_a$. Ohm's law applied to the membrane at the peak of excitation is given by

$$I = \frac{V - E_a}{r_a} \tag{2}$$

where r_a is the membrane resistance [see Adelman et al. (18)]. In the range where $V < E_a$, I is negative, i.e., inwardly directed across the membrane. The slope of the straight line is equal to the membrane conductance, $1/r_a$. By comparing the slopes of the two straight lines, one passing through V = O and the other through $V = E_a$, it is found that the membrane conductance at the peak of excitation is about 1/10 of the value in the resting state.

The state of the membrane carrying no net current has a special significance in the present discussion. Under the conditions that $V = E_a$ and I = O, the fluxes of calcium and sodium ions, denoted by $J_{Ca}^{(0)}$ and $J_{Na}^{(0)}$, respectively, and expressed in number of moles transported across unit area in unit time from inside to outside, are expected to satisfy

$$2 J_{Ca}^{(0)} + J_{Na}^{(0)} = 0 \tag{3}$$

(Note that the fluxes of anions are small in squid axons.)

When voltage V is only slightly different from E_a, current I varies directly with $(V - E_a)$ as indicated by Eq. (2). In this case, the relationship between the cation fluxes and the current is given by

$$I = F(2J_{Ca} + J_{Na}) \tag{4}$$

Subtraction of Eq (3) from (4) yields

$$I = 2F(J_{Ca} - J_{Ca}^{(0)}) + F(J_{NA} - J_{Na}^{(0)}) \tag{5}$$

This equation indicates that the membrane current consists of two components. The component corresponding to the first term on the right-hand side of the equation may be termed the "calcium current." The second component is what may be termed the "sodium current," corresponding to the last term in Eq (5). In other words, an inward membrane current under these bi-ionic conditions consists of a component representing an increase in the calcium influx and another component representing a decrease in the sodium efflux.

It is important to note that the argument stated above is valid inde-
pendently of the assumptions one has to make as to the mechanism of ion
transport across the membrane. There are strong fluxes of cations across
the membrane in the excited state of internally perfused squid giant axons.
Therefore, it is erroneous to ignore the second component in the membrane
current and to designate the total inward current as "calcium current."

Up to this point, the excited state of the axon membrane has been
treated as if it is a time-independent state. The excited state of the axon
under bi-ionic conditions lasts much longer than in unperfused axons. Reflecting this situation, the membrane current observed under these conditions varies slowly with time. Consequently, it is justifiable to treat the
axon membrane as being in a quasi-stationary state. Later on, in Sec.
VII, the time-dependence of the membrane potential and conductance is
discussed.

V. POTASSIUM ION AND THE RESTING POTENTIAL

Bernstein's membrane theory (19) postulates that the membrane potential of the axon in the resting state, E_r, is described by the following
equation

$$E_r = \frac{kT}{e} \ln \frac{[K]_i}{[K]_e} \tag{6}$$

where kT/e (= RT/F) has the usual physicochemical meaning (i.e., about
25 mV), and $[K]_e$ and $[K]_i$ are the external and internal potassium ion
concentrations, respectively. The concentration of potassium ion in sea
water is approximately 18 mM (20); that in the squid axoplasm is roughly
450 mM (21). The value of E_r calculated from these values is about 81 mV.
The observed value of the resting membrane potential, which is between 50
and 60 mV is definitely smaller than the calculated value.

By the use of the technique of intracellular perfusion, it is easy to
alter the internal potassium ion concentration over a wide range without
suppressing excitability. A ten fold dilution of the internal potassium salt
solution with an isotonic sugar solution was found to alter the resting potential by only about 10 mV instead of 58 mV as expected from Eq. (6)
(4, 22).

There is an alternative method of varying the internal potassium salt
concentration; that is to alter the ratio of $[K]_i$ and $[Na]_i$, maintaining the
total univalent cation concentration at a constant level (13, 23). An example
of the results of such alteration is shown in Fig. 6. It is seen in this figure

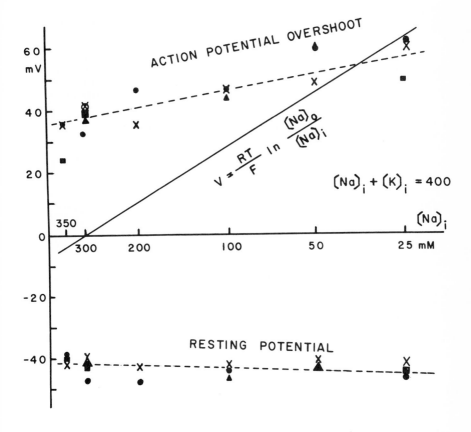

Fig. 6. Resting potential and overshoot of action potential of intra-cellularly perfused squid axons plotted as a function of Na concentration in the perfusing fluid, $[Na]_i$. The sum of the internal Na and K concentrations was held at a constant level of 400 meq/1 throughout. The perfusing fluid contained 470 mM glycerol besides sodium and potassium salts in the glutamate form. The external medium contained 300 mM NaC1, 45 mM $MgSO_4$, and 22 mM $CaCl_2$. A 50% increase in the Mg and Ca concentrations outside did not alter the results significantly. (From Tasaki and Takenaka, Proc. Natl. Acad. Sci. U.S., 50, 619, 1963.)

that, in a wide range of $[K]_i$ examined, the resting membrane potential remained practically unaffected by changes in $[K]_i$. It is also seen that in the entire concentration range, the axon maintained its ability to develop all-or-none action potentials. (There was a gradual fall in the action potential amplitude with increasing sodium ion concentration inside.)

It has long been known that a reduction in the potassium salt concentration in the external medium affects neither the resting potential nor the axon excitability significantly (2,4). Doubling the normal potassium ion concentration in sea water has very little or no affect on the membrane potential or on excitability of the axon immersed.

The conclusion that is drawn from the experimental facts stated above is clear: the axon membrane in its resting (and excitable) state does not possess an overwhelmingly high selectivity for potassium ion. (Note that a membrane is considered to have a high permeability specifically to one particular ion only when the membrane potential is governed predominantly by the particular ion in the presence of other ions at comparable or higher concentrations.)

The properties of the axon membrane become very different when the axon is depolarized by an excessive amount of potassium salt added to the external medium. Over the range of external potassium ion concentration between 50 and 500 mM, Eq (6) describes the observed changes in the membrane potential fairly well. It is important, however, to note that the axon is totally incapable of developing action potentials in a medium containing more than 50 mM potassium ion. Some physiologists believe that such depolarized axons are in their resting state, and this is the reason why Eq. (6) is believed to describe the resting potential.

In the following section, evidence will be presented indicating that the depolarized state of an axon is very different from its resting (and excitable) state in many respects.

VI. ABRUPT DEPOLARIZATION

In 1938, Osterhout and Hill (24,25) discovered that transition from the resting state of the Nitella membrane to its depolarized state is very abrupt. No physicochemical explanation was offered at the time of discovery as to the nature of this phenomenon of abrupt depolarization. Later on, essentially the same phenomenon was observed in nodes of Ranvier of the frog nerve fiber (26) and in squid giant axons (4,27).

Figure 7 shows an example of the experiments made on a squid giant axon internally perfused with a dilute cesium salt solution and immersed, initially, in a $CaCl_2$ (glycerol) solution. As has been stated earlier, the axon is electrically excitable under these conditions; but no electric current was delivered to the axon during the following observation:

The axon under observation was kept in a steady state by maintaining rapid circulation of both internal and external solutions. When the external

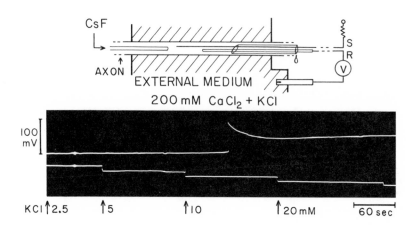

Fig. 7. Abrupt depolarization induced by external application of KC1
(upper oscillograph trace). The KC1 concentration in the rapidly flowing
external fluid medium was raised by a factor of two at the time marked by
the lower oscillograph trace. A silver wire electrode (S) was used to test
the excitability of the axon before the experiment. R represents a glass
pipet recording electrode. No electric stimuli were delivered while continuous recording of the membrane potential (V) was made. The potential
jump produced by 10 mM KC1 was 83 mV initially, and was followed by a
gradual potential fall of approximately 40 mV. Temperature, 16° C.
(From Tasaki, Takenaka, and Yamagishi, Am. J. Physiol., 215, 152,
1968.)

potassium ion concentration was raised from 0 to 2.5 mM (by replacing a
portion of glycerol with KC1), no significant change in the membrane potential was observed. Then, the potassium ion concentration was doubled;
once again, the observed change in the membrane potential was far smaller
than the values expected from Eq. (6). When the external potassium ion
concentration was doubled again, the membrane potential remained unaffected for a period of time; then, there was an abrupt rise in the intra-
cellularly recorded potential. The potential jump observed was about 83
mV; this jump was followed by a gradual potential fall of about 40 mV.

The potential jump observed under these conditions is far greater than
the value expected from Eq. (6) for doubling the external potassium ion
concentration. (The expected value is approximately 17 mV.) Once the
potential jump is observed, a further increase in the external potassium ion

concentration does not bring about any more sudden changes in the membrane potential. Instead, the potential smoothly changes with the potassium ion concention.

Abrupt depolarization strongly affects the electric excitability of the axon. Following an abrupt potential jump, the axon is no longer capable of developing action potentials in response to an applied outward membrane current. These observations are highly reproducible.

Measurements of the membrane impedance under these experimental conditions indicated that, at the moment when the membrane potential jumps, the membrane resistance suddenly falls (Fig. 8). From this and other tests, there seems little doubt that the observed potential jump represents initiation of an action potential, as originally suggested by Osterhout [see also (4, 28)]. The difference between an action potential and abrupt depolarization lies in the fact that an abrupt potential jump is not followed by repolarization (i.e., restoration of the original potential level) as ordinary action potentials do. According to this interpretation, a depolarized state of the membrane corresponds to the plateau stage of an action potential. In other words, the depolarized state logically comes under the category of the excited state, rather than the resting state.

A depolarized membrane is inexcitable when tested with a pulse of outwardly directed current. When tests are made with pulses of inward current, however, a distinct sign of excitability can be observed [Tasaki (26), Segal (29), Stampfli (30)]. Electric responses observed under these conditions are reversed in sign.

When the axon membrane is in the depolarized state, lowering of the external potassium ion concentration produces repolarization; this process of repolarization is often abrupt. When the external potassium ion concentration is only slightly above the critical concentration required for abrupt depolarization, an increase in the external calcium ion concentration (without changing the potassium ion concentration) brings about repolarization. Evidently, this is an example of the potassium-calcium antagonism well known to physiologists for many years [see Loeb, (31)].

The conductance of membrane with negative fixed charges is determined by the mobilities and the concentrations of the cations within the membrane. Hence, a sudden increase in the membrane conductance is a reflection of a sudden increase in the density of the negative sites in the membrane and/or a sudden decrease in the compactness of the membrane. It is reasonable to assume that such a structural change in the membrane is caused by replacement of calcium ions within the membrane by potassium ions. The fact that a continuous rise in the external potassium ion

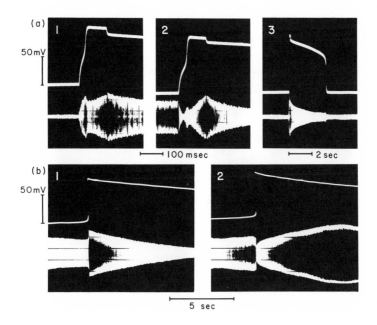

Fig. 8. (a): Changes in the membrane impedance associated with electrically induced action potential of an axon internally perfused with 25 mM CsF solution and immersed in 200 mM CaCl₂ solution. The impedance bridge was balanced with the membrane impedance either in the resting state [records (a), 1 and (a), 3] or during excitation (record (a), 2). The room temperature was 18° C. (b): Changes in the membrane impedance produced by a gradual alteration of the composition of the external medium from a mixture of 200 mM CaCl₂ and 25 mM CsC1 to a mixture of 450 mM CsC1 and 100 mM CaCl₂. The two records, (b), 1 and (b), 2 were taken from the same axon. After record (b), 1 was taken, the medium was immediately switched to the CaCl₂-rich mixture; the impedance bridge was then readjusted before another application of the cesium-rich solution [record (b), 2]. Note that the sudden rises of the membrane potential were associated with simultaneous changes in the membrane impedance. Temperature, 19° C. (From Tasaki, Takenaka, and Yamagishi, Am. J. Physiol., 215, 152, 1968.)

concentration produces an abrupt change in the membrane structure is a sign of cooperativity of the process involved [see also Changeaux et al. (32), Hill (33), Lehninger (34), Adam (35), Lissowski (36)].

   More recently, it was found that abrupt depolarization can be evoked by a number of univalent cations other than potassium ion. However, the

critical concentration required for abrupt depolarization is very different
for different cation species. Among common univalent cations, the ion
sequence arranged in accordance with the ability to depolarize the membrane is

$$K > Rb > NH_4 > Na, Li$$

where potassium ion has the strongest depolarizing power, i.e., the lowest
critical concentration for depolarization. Since abrupt depolarization is a
process occurring at the external layer of the membrane, the difference in
the chemical species of the internal univalent cation does not influence the
ion sequence for depolarization.

It is very interesting to note that the ion sequence described above is
very similar to that determined by several investigators for the so-called
sodium-pump mechanism [see e.g. (37)]. The reason for this similarity
may be found in the mechanism of ion-accumulation proposed by Tasaki and
Singer (9) some time ago. According to this mechanism, particular cations
are driven into the axoplasm as the result of rhythmical appearance of
randomly distributed depolarized spots in the membrane. Consequently,
cations with strong depolarizing power are preferentially accumulated in the
axoplasm. It seems reasonable that phosphoproteins in the membrane play
an essential role in the process of ion-accumulation [see e.g. Heald (38)].

VII. TWO STABLE STATES OF THE MEMBRANE

It is shown in the preceding section that an axon can be brought abruptly to a depolarized state by slowly increasing the external univalent cation
concentration. One might argue that the process of abrupt depolarization
could be treated as one form of excitation by electric current. It is reasonable to assume that concentration of the externally applied potassium or
cesium ion is not perfectly uniform along the axon. Such nonuniformity
could generate electric currents traversing the axon membrane inwards in
the highly depolarized zones and outwards in the weakly depolarized or normal zones. Since an outward membrane current is capable of evoking an
action potential, abrupt depolarization could be attributed to initiation of an
action potential by this mechanism.

There is, however, serious difficulty in this apparently reasonable
interpretation. Experiments with both Nitella and internally perfused squid
axon indicate that there is a distinct discontinuity in the steady-state values
of the membrane potential when plotted against the external potassium or
cesium ion concentration. There is also a large discontinuity in the steady-
state values of the membrane conductance. Since the action potential is

defined as a transient change in these quantities, it is not possible to attribute the observed potential and conductance changes simply to production of an action potential. Thus, we may conclude that the phenomenon of abrupt depolarization cannot be explained in terms of the classical concepts in electrophysiology (without introducing serious modifications or ad hoc assumptions).

Traditionally, physiologists turn to physical chemistry for explanations whenever they are confronted with a new phenomenon. We find that the difficulty of explaining the phenomenon under study can be resolved immediately if we introduce the concept of cooperative cation-exchange process.

Experimental evidence for the existence of fixed negative charges in the outer layer of the axon membrane has been discussed earlier. The density of the fixed charge is shown to be high enough to exclude anions (derived from the external medium) almost completely. When a giant axon is immersed in a $CaCl_2$ (glycerol) solution, the anionic sites in the outer layer of the axon membrane are occupied almost exclusively by calcium ions. We now add a small amount of KCl to the external medium; then, a portion of the calcium ions at the anionic sites are replaced with potassium ions. As the external potassium ion concentration is raised continuously, this potassium-calcium exchange process continues.

We now assume that this ion-exchange process involves a cooperative, discontinuous step. Figure 9 shows a theoretical ion-exchange isotherm for an ion exchanger with such cooperative characteristics. We assume that this isotherm describes the behavior of the outer layer of the axon membrane. The abscissa in this diagram represents the fraction of the univalent cations in the external medium; the ordinate represents the fraction of the anionic sites in the membrane occupied by univalent cations. When the external univalent cation concentration reaches the level marked by C in the figure, there is a sudden increase in the fraction of the univalent cation in the membrane; this discontinuous step is represented by the broken line C---E. A further increase in the external univalent cation concentration increases the intramembrane univalent cation concentration continuously (see line E-F).

The selectivity of the membrane for the univalent cation over the divalent cation, K, is defined by the formula

$$\frac{\bar{C}_1^2}{\bar{C}_2} = K \frac{C_1^2}{C_2}$$

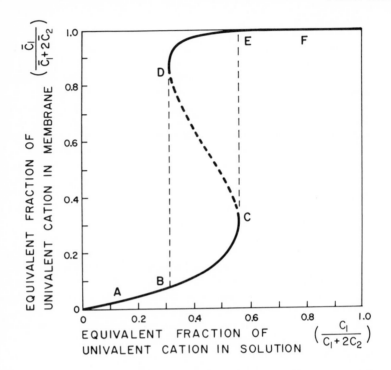

Fig. 9. Theoretical ion-exchange isotherm calculated for a cation-exchanger membrane immersed in a salt solution containing univalent and divalent cations. In calculation, it was assumed that occupancy of two neighboring charge sites in the membrane by two cations of different valences is energetically unfavorable. C_1 and C_2 represent the concentrations in the solution of univalent and divalent cations respectively. $\overline{C}_1$ and $\overline{C}_2$ represent the concentrations within the membrane. When the equivalent fraction of the univalent cation in the solution is increased continuously from zero to one, the corresponding fraction in the membrane increases along the course, O, A, B, C, E, and F. When the equivalent fraction in the solution is decreased from one to zero, the corresponding fraction in the membrane changes along F, E, D, B, A, O. (Adapted from Tasaki J. Gen. Physiol., 46, 755, 1963.)

where C_1 and C_2 represent the univalent and divalent cation concentration in the external medium and $\overline{C}_1$ and $\overline{C}_2$ are the values in the outer layer of the axon membrane, respectively [see e.g., p. 153 in Helfferich (17)]. Reflecting the situation that two univalent ions are needed to replace one divalent ion, squares of the univalent cation concentrations appear in the equation. As C_1 varies along the line O-A-B-C, the selectivity, K, varies continuously. When the intramembrane univalent cation concentration, $\overline{C}_1$, suddenly increases (C --- E), there is a sudden rise in the selectivity. A further increase in C_1 raises the selectivity continuously.

A sudden increase in the selectivity for the univalent cation at the outer membrane layer brings about a sudden rise in the intracellular potential. Since the intramembrane mobility of univalent cations is greater than that of divalent cations, a sudden rise in the intracellular potential is accompanied by an abrupt rise in the membrane conductance. Thus, the phenomenon of abrupt depolarization is adequately explained on the basis of the cooperative ion-exchange process.

The isotherm shown in Fig. 9 predicts that a hysteresis loop may be observed in this cation-exchange process. When the external univalent cation concentration, C_1, is lowered continuously from the point marked F, a sudden fall in the selectivity is expected to proceed along the line marked D---B, and not along E---C. A further decrease in $C_1/(C_1 + 2C_2)$ alters the selectivity continuously. This prediction was verified by a series of experiments conducted on internally perfused squid giant axons (27). It was found that both the membrane potential and conductance return to the lower values at a concentration far lower than the value at which abrupt depolarization is produced when C_1 is raised slowly.

From these discussions, it is found that the portion of the isotherm shown by A-B-C represents the resting (repolarized) state of the axon and portion D-E-F represents the excited (depolarized) state. In the range of the external conditions marked by vertical line B-D and line C-E, there are two possible states which the membrane can assume. These two states are stable (except at point C and D), in the sense that small perturbations do not trigger large changes in the properties of the membrane and that the membrane returns to the original state soon after the perturbations are removed. For this reason, the logical framework described in this section is called the "two-stable-states hypothesis."

The ion-exchange isotherm shown in Fig. 9 is derived under the assumption that occupancy of neighboring sites by cations of the same kind is energetically more favorable than occupancy of adjacent sites by different cation species. This assumption has been examined extensively by Adam (35) and by Loussowski (36).

Finally, we discuss the time-dependent part of the membrane potential and conductance seen in Figs. 7 and 8. The ion-exchange isotherm shown in Fig. 9 does not, by itself, explain how the potential gradually changes following transition. However, the following consideration suggests that the time-dependence of the membrane potential may be the consequence of the decrease in membrane resistance in the depolarized state.

In the resting state, the membrane resistance is high; therefore, the
fluxes of cations across the membrane are small. When the membrane
resistance suddenly falls following transition, there is a corresponding rise
in the cation fluxes. (In accordance with the Nernst-Einstein relationship,
there is inverse proportionality between the fluxes and the resistance; see
Kobatake and Tasaki in Ref. 4.) This rise in the cation fluxes gradually
decreases the concentration difference (across the membrane) of the cations
contributing to the membrane potential, resulting in a gradual shift of the
potential.

There seems little doubt that the major factor which causes the gradual
fall in the membrane potential is the increased interdiffusion of cations described above. However, some other type of relaxation phenomenon may
make some additional contribution to the potential fall.

VIII. ELECTRIC STIMULATION

Let us consider an axon internally perfused with a solution of the salt
of univalent cations (say cesium) and immersed in a mixture of calcium
salt and the salt of the same univalent cations (cesium). When the external
concentration of univalent cation is within a certain limited range, the axon
is electrically excitable. Under these conditions, it is not difficult to understand the mechanism of eliciting an action potential by a pulse of electric current on the basis of the ion-exchange isotherm shown in Fig. 9.

Let us suppose that the condition of the external medium corresponds
to the portion of the isotherm indicated by A--B; the axon is initially in the
resting state. Now, we deliver a pulse of outward current which tends to
transport the internal cesium ions into the outer layer of the axon membrane. Since the occupation of neighboring anionic sites by the same cation
species is considered to be energetically favorable, cesium ions tend to
cluster in the outer layer of the axon membrane and to form patches or
spots which are in the active (i. e. , electrophysiologically excited) state.
Hence, a pulse of outward current is expected to create additional "active
spots" in the membrane. When the current is strong enough and its duration long enough, an abrupt jump in ion-selectivity (step C--E in Fig. 9) is
expected to take place, resulting in initiation of an action potential.

[The critical level of the membrane potential at which an all-or-none
action potential is released is usually 15 to 35 mV. Some investigators
stress the importance of the electric field strength produced by this applied
potential (I-R) drop. But, it is to be noted that this potential drop is close
to the value of kT/e (= 25 mV at room temp). This value may be regarded
as the unit of potential in Nernst-Planck flux equation. In a spatially uniform

membrane, this potential drop changes the cation fluxes under the bi-ionic condition by only about 50% (see Appendix III in Ref. 4). Therefore without the assumption of cooperativity and spatial nonuniformity of the membrane, it seems extremely difficult to understand how a potential drop of about 25 mV generates an action potential.]

We now discuss the mechanism of termination of an action potential. The results of impedance measurements on axons under bi-ionic conditions (Fig. 8) indicate that the membrane resistance progressively increases during action potential. From this increase, it follows that the fluxes of cations through the membrane (i.e., the rate of arrival of univalent cations from the axon interior) gradually fall while the axon is in the active state. This fall of cation fluxes in turn raises the divalent cation concentration in the outer membrane layer. Eventually, the membrane undergoes transition D---B (in Fig. 9), bringing about an abrupt repolarization.

According to the present hypothesis, the downward transition in the state of the membrane (D---B) is in many respects analogous to the upward transition (C---E). When an axon is in the state marked D-E, a pulse of inward current can produce a transition to a repolarized state. Under the bi-ionic conditions, we have frequently encountered axons which spontaneously oscillate between the depolarized and repolarized state. By adjusting the univalent-to-divalent cation concentration ratio in the medium, it is often possible to demonstrate action potentials of the reversed sign, namely, a transition D--B caused by a brief current pulse followed by a spontaneous depolarization (i.e., gradual and sudden changes in the state of the membrane along the line B--C--E--F).

It is interesting to introduce, after Changeux et al. (32) and Lehninger (34), the concept of the "membrane subunit." Although the existence of such a subunit can not be proven directly under physiological conditions, it is reasonable to assume that a small but finite portion of the membrane has to be altered by the ion-exchange process to form an "active spot." Undoubtedly, the smallest size of an active spot is determined by the size of individual membrane macromolecules and the range of intermolecular forces. Figure 10, taken from Lehninger's article, shows various conformational states of the axon membrane. The term "stimulus" in the figure may be taken either as the univalent cation fraction (as in Fig. 9) or as the strength of an outwardly directed electric current. The term "response" may be regarded as representing either the membrane potential or the membrane conductance. The four states of the membrane shown in this diagram may be compared to states A, C, E, and F in Fig. 9.

The mechanism of excitation of an axon under multi-ionic conditions can also be explained on the basis of the concept illustrated in Figs. 9 and

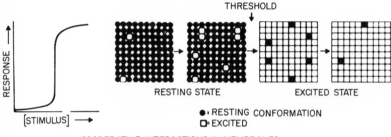

Fig. 10. Diagram showing cooperative interactions in membrane structure. (From Lehninger, Proc. Nat. Acad. Sci. U.S., 60, 1069, 1968.)

10. When the major cation species in the external medium are sodium and calcium, the axon membrane can be in the resting (i.e., repolarized) state even at a relatively high sodium ion concentration. (Note that the depolarizing power of sodium ion is very low.) If potassium ion is the major cation species in the axon interior, a pulse of outward current delivered through the axon membrane initiates invasion of potassium ions into the outer membrane layer. Because of the strong depolarizing power of potassium ions, this invasion creates a rapid increase in the number of active spots. When the critical number of active spots is reached, an abrupt transition to the depolarized state is produced.

A simple calculation, based on the difference in membrane potential and conductance between active and resting states of the membrane [p. 133 in Ref. (4)] indicates that the critical active fraction is about 1% or less of the total membrane. This means that, when an area of only 1% or less of the total membrane is brought into the active state, the entire membrane subunits become predominantly active. The rate of the potential fall from the peak of the action potential is high under these conditions. This high rate is attributed to the high membrane conductance (creating strong interdiffusion of cations) and the high selectivity of the active membrane to potassium ions (13).

IX. MEMBRANE POTENTIAL AND EQUIVALENT CIRCUIT

In the theory of nerve excitation widely accepted at present (1, 2), the axon membrane is represented by an equivalent circuit consisting of two batteries with fixed emf's (sodium and potassium batteries), two variable

resistors, and a capacitor. In this circuit, the membrane potential is determined by the electric charge on the capacitor. For this reason, it is
stated frequently that inwardly directed transport of cations across the
membrane (notably sodium influx) is responsible for production of action
potentials. Obviously, this mechanism of nerve excitation is very different
from the macromolecular approach described in preceding sections.

It is important to note that the concept of equivalent circuits is not used
in physicochemical theories of inanimate membranes. Physical chemists
construct their theories based on thermodynamics, whereas axonologists
develop their ideas along the concepts used by electrical engineers. The
electrical engineers' approach may, in some instances, yield the same
result as a thermodynamic approach does. In general, however, two different approaches are expected to lead to different conclusions.

A physicochemical treatment of a uniform membrane with a high fixed
density of negative charge indicates that the membrane potential observed
under the bi-ionic conditions is given by the following formula

$$\varphi = \ln \frac{(C_2 X)^{\frac{1}{2}}}{C_1} - \frac{1}{2}\ln 2K + \frac{u_1 - u_2}{u_1 - 2u_2} \ln \frac{2u_2}{u_1}$$

where φ is the membrane potential in the unit of kT/e (= 25 mV), C_1 and C_2
are the concentrations of the univalent and divalent cation concentration,
respectively, X is the fixed charge density, K is the selectivity for the
univalent cation over the divalent cation, and u_1 and u_2 are the mobilities
of these cations [see p. 56 in Ref. (4)]. Under the conditions in which C_1
and C_2 are given, the membrane potential varies when one or more of the
three quantities, X, K, and u_1/u_2, are altered. In the preceding section,
we have attributed abrupt changes in the membrane potential to sudden
alteration in these quantities associated with phase transition. In this
physicochemical approach, it is not required to introduce batteries with
fixed emf's and separate resistors to describe the behavior of the membrane.

It is well known that a strong pulse of outward-directed current excites
the axon membrane in an extremely short period of time. When the stimulating pulse is about five times the threshold in strength and roughly 50-to
100 μsec in duration, the entire stimulated area of the axon membrane is
expected to be thrown into the excited state before the end of the applied
current pulse. Obviously, there is no inward current across the membrane
at any time during this process of evoking an action potential. Therefore,
it is clear that an inward (sodium) current is not a necessary antecedent
of an action potential.

If the stimulating pulse is not strong enough, only a portion of the total area of the membrane may be brought into the excited state at the moment when the applied current pulse is terminated. In this case, local (or eddy) currents are set up in the membrane after the end of the applied pulse. The local currents are outwardly directed across the resting portion of the membrane and inwardly directed in the excited portion. The inward current tends to induce abrupt repolarization in the excited portion of the membrane and the outward current tends to bring about abrupt depolarization in the resting portion. As the consequence of this long range interaction between the active and resting portions of the membrane, the entire axon membrane approaches either one of the two stationary (non-equilibrium) states, resting or active. In summary, it is outward membrane current -- and not an inward current -- that causes a transition of the membrane from the resting to the active state.

Finally, a comment is made on the two separate resistors in the equivalent circuit. An electric current traversing a cation-exchanger membrane is carried, as is well known, predominantly by cations within the membrane. It is illogical and incorrect, however, to regard the electric conductance of the membrane as consisting of separated conductances of individual cations. (The only exception to this statement is the case in which the solutions on both sides of the membrane have identical or nearly identical compositions.) If membrane conductance is measured with a high frequency ac, the fraction of the current carried by one particular cation species varies from one layer to another within the membrane; hence, it is absurd, under these conditions, to represent the membrane with separate conductors for individual cations connected in parallel. Even when a weak steady current is used to measure the membrane conductance, the total conductance cannot be given by the sum of contributions of individual cation species [see pp. 41-42 in Ref. (4)].

The success of the equivalent circuit theory in deriving the time course of the action potential from the two time-dependent conductances does not by itself guarantee the legitimacy of separation of the membrane conductance into two components. According to the two-stable-states hypothesis, the membrane currents observed under voltage-clamp conditions are nothing more than action currents distorted by the clamping device. It is not surprising that one can derive the time course of the action potential (observed in the absence of current) from a large number of action current records.

X. OPTICAL CHANGES IN AXON MEMBRANE DURING ACTION POTENTIAL

Until quite recently, rapid physicochemical changes occurring during action potential in and near the axon membrane could be detected only by

electrophysiological techniques. The main difficulty in applying various
optical techniques to follow these rapid changes derives from the situation
that the axolemma is extremely thin.

At present, it is known that (a) light scattering, (b) birefringence, and
(c) extrinsic fluorescence of axons undergo transient changes during action
potential (39-42). In this article we are concerned only with changes in
extrinsic fluorescence, because the data obtained from fluorescence measurements are readily amenable to interpretations on the molecular basis.

Figure 11 shows the principle of the technique used and an example
of the optical records obtained from a crab nerve stained with acridine
orange. The white light from a strong incandescent lamp was converted
into quasi-monochromatic light of 465 m$_\mu$ by inserting an interference filter
(F$_1$ in the figure) between the light source and the nerve. By the use of
lenses, the light was focused on the stained nerve. The fluorescent light
from the nerve was detected with a photomultiplier tube used in conjunction
with a secondary filter (F$_2$) which interrupted the scattered incident light
and passed the emitted light. The nerve was stimulated with brief pulses
of electric current applied to the nerve near its end; the action potential
was recorded extracellularly with a pair of electrodes placed on the surface
of the nerve at the other end.

It is seen in the figure that an action potential evoked under these conditions is accompanied by a simultaneous increase in the intensity of the
fluorescent light. The increase was (2 to 5) x 10^{-4} - times the intensity of
the fluorescent light before stimulation. The time course of the optical
signal was similar to that of the action potential. In a series of experiments with squid giant axons, acridine orange was introduced into the axon
interior by the use of the technique of intracellular perfusion. There was
an increase in the fluorescent light when the axon was stimulated under
these conditions.

In a squid giant axon with acridine orange internally and sea water
externally, the dye molecules are distributed in the axoplasm (which does
not participate in action potential) as well as in the axon membrane. Therefore, it is clear that the ratio of change in fluorescence occurring in and near
the membrane to background intensity is far greater than the values found
in these experiments. Acridine orange molecules are known to be bound to
negatively charged sites in various macromolecules (43). For this and
other reasons, the dye molecules are believed to be present in the outer
layer of the axon membrane. The quantum yield of fluorescence is very
sensitive to the microenvironment of the dye molecules (44). It seems
reasonable to suggest that the fluorescence increase is due to a reduction
of self-quenching associated with formation of new negative sites during
action potential.

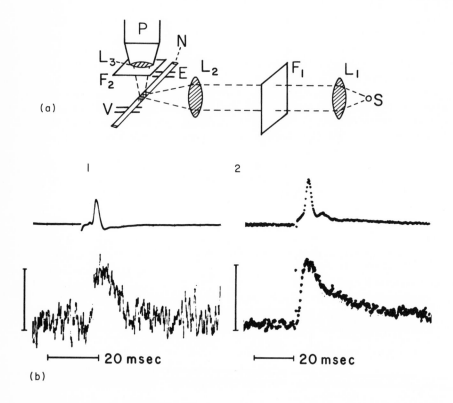

Fig. 11. (a): Schematic diagram of the experimental arrangement used for the detection of fluorescence changes associated with nerve conduction. The letter S represents the light source; L_1, L_2, and L_3, lenses; F_1 and F_2, optical filters; N, nerve; E and V, stimulating and recording electrodes, respectively; and P, photomultiplier tube. (b): Electric (top) and optical (bottom) responses obtained from crab nerves stained with acridine orange; the vertical bars indicate an increase of 5×10^{-4} (record 1) and 2.5×10^{-4} (record 2) times the background intensity. In record, the noise level of the optical signal was reduced by the use of a CAT computer. Temperature was 19° C. (From Tasaki, Carnay, Sandlin, and Watanabe, Science, 163, 683, 1969.)

At present, the following fluorescent compounds are known to give rise to optical signals: acridine orange, acridine yellow, 8-anilinonaphthalene-1-sulfonate (ANS), auramine 0, 1-dimethylaminonaphthalene-5-sulfonyl chloride (DNS), fluorescein isothiocyanate, LSD, pyronin B, pyronin Y, rhodamine B, rhodamine G, rose bengal, and 2-toluidinylnaphthalene-6-sulfonate (TNS), etc. It is to be noted that some of these dyes give rise to

negative signals (representing transient decrease in fluorescence) during
action potential (42).

It is well known that measurements of fluorescence polarization offer
direct information about the viscosity of the medium surrounding fluorescent
molecules. In crab nerve stained with pyronin B, the degree of polarization of the fluorescent light was found to decrease following nerve stimulation. From this finding it follows inferentially that the intramembrane
viscosity falls at the time when an action potential is produced. Figure 12,
top, illustrates the principle of the method used to obtain these results.

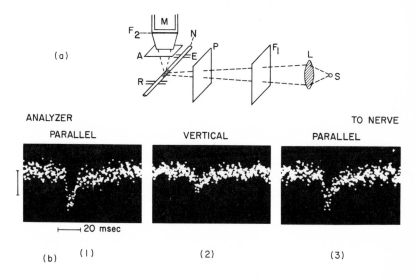

Fig. 12. (a): Schematic diagram showing the experimental setup used
to measure changes in fluorescence polarization associated with production
of action potential in a crab nerve stained with pyronin B. The letter S
represents the light source; L, lens; F_1, primary filter; F_2, secondary
filter; M, multiplier phototube; P, polarizer; A, analyzer; N, nerve; E,
stimulating electrodes; V, recording electrodes. (b): Optical signals obtained with the principal axis of the analyzer (E-vector of fluorescent light)
parallel to the long axis of the nerve [records 1 and 3] and perpendicular
to the nerve [record (2)]. The quasi-monochromatic light used for excitation of dye molecules was polarized (with its E-vector) in the direction
parallel to the nerve. A CAT computer was used for recording. For records 1 and 3, the vertical bar represents 2×10^{-4} times the background
intensity. Record 2 was taken under the same experimental conditions
except that the analyzer was rotated by $90°$. Temperature, $19°$ C. (Adapted from Tasaki, Carnay and Watanabe, Proc. Natl. Acad. Sci. U.S.,
64, 1362, 1969.)

Basically, the experimental setup is similar to that shown in Fig. 11. In addition to the source of quasi-monochromatic light and the detector of fluorescent light, there is a polarizer between the light source and the nerve and an analyzer between the nerve and the detector.

The pyronin B molecules in the nerve were excited with a quasi-monochromatic light (550 mμ) polarized in the longitudinal direction of the nerve (i.e., the electric field vector parallel to the nerve). Changes in the fluorescent light from the nerve was measured initially with the analyzer (E-vector of the emitted light) parallel to the nerve (Record 1). Then, the analyzer was rotated by 90° and a measurement was made of the fluorescence changes during action potential (Record 2). By comparing the optical signals obtained under these two conditions, it was found that the signal with the analyzer parallel to the nerve is approximately twice as large as that obtained with the analyzer vertical to the nerve.

The degree of polarization is defined as

$$P = \frac{I' - I''}{I' + I''}$$

where I' and I'' are the intensity of the fluorescent light polarized longitudinally and vertically, respectively, relative to the nerve; P can also be expressed in terms of polarization ratio, R:

$$P = \frac{1 - R}{1 + R}$$

where $R = I''/I'$.

In nerves stained with pyronin B, R was approximately 0.75 in the resting state. During action potential, the intensity of the fluorescent light decreases; I' decreases much more than I'' does. This means that the polarization ratio, R, increases and the degree of polarization, P, decreases during action potential.

According to the theory developed by Perrin (45), a decrease in the degree of polarization of the fluorescent light is the direct consequence of increased rotational Brownian movement due to a decrease in the viscosity of the medium surrounding fluorescent molecules. This experimental finding, suggesting a transient fall in the intramembrane viscosity, has offered a new basis for physicochemical interpretation of the process of nerve excitation. There seems little doubt that the branch of molecular biology dealing with the problem of nerve excitation will make rapid progress in the future by the use of various optical techniques.

XI. SUMMARY AND CONCLUSION

The technique of intracellular perfusion offers a means of studying various properties of the axon membrane under well defined chemical environments. Consequently, the experimental results obtained by the use of this technique are amenable to physiochemical interpretations.

Recent studies of the role of sodium ion in the process of action potential production indicates that the presence of this ion is not a prerequisite to nerve excitation in squid giant axons. Axons are found to maintain excitability in external media completely devoid of univalent cations. All-or-none action potentials of about 60 mV in amplitude are observed in axons with sodium-phosphate internally and $CaCl_2$ as the sole external electrolyte. From these observations it is concluded that the external calcium ion (but not sodium ion) plays an essential role in nerve excitation.

Electrophysiological evidence for the existence of a cooperative ion-exchange process involving univalent and divalent cations in the membrane is discussed. The mechanism of nerve excitation is explained on the assumption that occupation of neighboring anionic sites in the membrane by the same cation species is energetically more favorable than occupation by different cation species.

Recent observations on changes in extrinsic fluorescence in nerve membrane during action potential are described. By the method of measuring the degree of polarization during action potential, possible changes in intramembrane viscosity are demonstrated.

ACKNOWLEDGMENTS

I wish to thank Dr. Akira Watanabe, Dr. Lawrence Lerman, and Dr. Laurence Carnay for their valuable assistance in conducting experiments described in this article.

REFERENCES

1. K. S. Cole, Membranes, Ions and Impulses, Univ. of California, 1968.
2. A. L. Hodgkin, The Conduction of the Nervous Impulse, Liverpool University, 1964.
3. A. L. Hodgkin, and B. Katz, J. Physiol. (London), 108, 37 (1949).
4. I. Tasaki, Nerve Excitation: A Macromolecular Approach, Charles C Thomas, Springfield, Ill., 1968.
5. H. S. Frank, and W. Y. Wen, Discussions Faraday Soc. 24, 133-141 (1957).

6. F. J. Millero, "The Partial Molal Volumes of Ions in Water" in Structure and Transport Processes in Water and Aqueous Solutions, (R.A. Horne, ed.), Wiley 1969.

7. R. de Nó Lorente, F. Vidal and L.M.H. Larramendi, Nature, 179, 737 (1957).

8. I. Tasaki, I. Singer, and A. Watanabe, Proc. Natl. Acad. Sci. U.S., 54, 763 (1965).

9. I. Tasaki and I. Singer, Ann. N.Y. Acad. Sci., 137, 792 (1966).

10. I. Tasaki, A. Watanabe, and I. Singer, Proc. Natl. Acad. Sci. U.S., 56, 1116 (1966)

11. I. Tasaki, A. Watanabe, and L. Lerman, Am. J. Physiol., 213, 1465 (1967).

12. A. Watanabe, I. Tasaki, and L. Lerman, Proc. Natl. Acad. Sci. U.S. 2246 (1967).

13. I. Tasaki, L. Lerman, and A. Watanabe, Am. J. Physiol., 216, 130 (1969).

14. I. Tasaki, I. Singer, and T. Takenaka, J. Gen. Physiol., 48, 1095 (1965)

15. T. Teorell, and C.S. Spyropoulos, cited on p. 57 in ref. 4.

16. K. Sollner, S. Dray, E. Grim, and R. Neihof, Ion Transport Across Membranes, (H.T. Clark, ed.), Academic, New York 1954, p. 144.

17. F. Helfferich, Ion Exchange, McGraw-Hill, New York, 1962.

18. W.J. Adelman, F.M. Dyro, and I. Senft, J. Gen. Physiol., 48, supple. 1 (1965).

19. J. Bernstein, Electriobiologie, Braunschweig, Fr. Vieweg, 1912.

20. G. M. Cavanaugh (ed.), Formulae and Methods IV of the Marine Biological Laboratory, Woods Hole, Mass., 1956.

21. H. B. Steinbach, and S. Spiegelman, J. Cellular Comp. Physiol., 22, 187 (1943).

22. P. F. Baker, A. L. Hodgkin, and T. Shaw, J. Physiol. (London), 164, 355 (1962).

23. I. Tasaki, and T. Takenaka, The Cellular Functions of Membrane Transport, (J. E. Hoffman, ed.), Englewood Cliffs, Prentice-Hall, 1964, p. 95.

24. W. J. V. Osterhout, and S. E. Hill, J. Gen. Physiol., 22, 139 (1938).

25. S. E. Hill, and W. J. V. Osterhout, J. Gen. Physiol., 21, 541 (1938).

26. I. Tasaki, J. Physiol. (London), 148, 306 (1959).

27. I. Tasaki, T. Takenaka, and S. Yamagishi, Am. J. Physiol., 215, 152 (1968).

28. J. P. Reuben, P. W. Brandt, L. Girardier, and H. Grundfest, Science, 155, 1263 (1967).

29. J. R. Segal, Nature, 182, 1370 (1958).

30. R. Stämpfli, Helv. Physiol. Pharmacal. Acta, 16, 127 (1957).

31. J. Loeb, The Dynamics of Living Matter, Columbia Univ. New York, 1906.

32. J-P. Changeux, J. Thiery, Y. Tung, and C. Kittel, Proc. Natl. Acad. Sci. 57, 335 (1967).

33. T. L. Hill, Proc. Natl. Acad. Sci. U.S. 58, 111 (1967).

34. A. L. Lehninger, Proc. Natl. Acad. Sci. U.S., 60, 1069 (1968).

35. G. Adam, Physical Principles of Biological Membranes, Gordon & Breach Science Publisher, New York 1970.

36. A. Lissowski, Proc. Third Intern. Biophys. Congr., II-X-8, 1969.

37. P. F. Baker, M. P. Blaustein, R. D. Keynes, J. Manil, T. I. Shaw, and R. A. Steinhardt, J. Physiol. (London), 200, 459 (1969).

38. P. J. Heald, Nature, 193, 451 (1962).

39. L. B. Cohen, R. D. Keynes, and B. Hille, Nature, 218, 438 (1968).

40. I. Tasaki, A. Watanabe, R. Sandlin, and L. Carnay, Proc. Natl. Acad. Sci. U.S., 61, 883 (1968).

41. I. Tasaki, L. Carnay, and A. Watanabe, Science, 163, 683 (1969).

42. I. Tasaki, L. Carnay, and A. Watanabe, Proc. Natl. Acad. Sci. U.S., 64, 1362 (1969).

43. D. F. Bradley, and M. K. Wolf, Proc. Natl. Acad. Sci. U.S., 45, 944 (1959).

44. R. W. Albers, and G. J. Koval, J. Biochim. Biophys. Acta, 60, 359 (1962).

45. F. Perrin, Compt. Rend. Acad. Sci., 180, 581 (1925).

NOTE ADDED IN PROOF

More recent work on extrinsic fluorescence using the dye 2-p-toluidinyl-6-naphthalene-sulfonate supports the concept of a hydrophobic-hydrophilic transition of the membrane during the action potential and suggests that the membrane has a crystalline structure [I. Tasaki, A. Watanabe, and M. Hallett, Proc. Natl. Acad. Sci. U.S.A., 68, 938 (1971)].

Chapter 4

DIVALENT CATION, ORGANIC CATION, AND POLYCATION
INTERACTION WITH EXCITABLE, THIN LIPID MEMBRANES

Ross C. Bean, William C. Shepherd, and Joellen T. Eichner

Philco-Ford Corporation
Aeronutronic Division
Newport Beach, California 92663

I. INTRODUCTION

Despite a long-term realization of the important participation of
divalent cations (magnesium and calcium) and a number of organic cations
in permeability control of cellular membranes, their activities are still
poorly understood. This is partly due to the complexity of the cell and cell
membrane, which has so far prevented determining the nature of the substances with which these ions interact; and partly to the diversity of the
activities of the ions themselves. Calcium or magnesium will produce a
high resistance in a normally low resistance junctional wall of an epethelial
cell (1) or stabilize the high resistance, resting state of the excitable cell
(2) They may be the transport ion of the action potential in crustacean
muscle (3) and are required in the active transport of other ions (4). Magnesium and calcium ion activities are also tightly interwoven with the
numerous permeability related functions of physiologically active amines
(5). Many attempts have been made to circumvent the difficulties inherent
in the direct studies on cells through utilization of various model systems
to elucidate and define the nature of cation interactions with the membrane
structure. Some important information has been gained on the reactions
which inorganic and organic cations may undergo with membrane components in experiments with lipid monolayers at aqueous-air (6, 7) or
mercury-water (8) interfaces; in complex, polymer supported lipid membranes (9, 10); or various films prepared from synthetic or biological
materials (11).

Among the many synthetic models, only one, the "excitable lipid
bilayer" membrane (12-16), has been able to produce a variable conductance with a negative differential resistance, characteristic of the excitable
cellular membrane still differ markedly in a number of respects, they bear
of divalent ions found in cellular membranes. While this model and the
cellur membrane still differ markedly in a number of respects, they bear
sufficient resemblance, both structurally and behaviorally, to make it
worthwhile to review here some of the model membrane properties -- in
an effort to develop some insight into possible mechanisms for control of
ion permeability in membranes through divalent cation or organic cation
interaction. To provide an adequate base for understanding the cation
control reactions, some basic properties of the lipid bilayer membrane
and the excitable lipid bilayer membrane must be presented in some detail.
It may then be seen that the sometimes puzzling action of the controlling
cations follows a reasonable pattern.

II. THE LIPID BILAYER MEMBRANE: GENERAL DESCRIPTION

A. Formation

Mueller and co-workers (12, 13) first developed a reliable procedure
for generating a very thin lipid membrane under water, stretched across

an aperture in a septum separating two aqueous compartments (Fig. 1). This was accomplished simply by stroking a small brush, loaded with an amphipathic lipid solution, across the aperture to form a relatively thick film (0.2-5 μm) which spontaneously thinned to a black (nonreflective) film. Numerous measurements (see below) indicate that the black film is normally less than 100 Å in thickness, corresponding to the dimensions normally

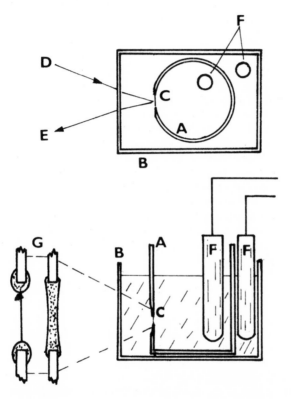

Fig. 1. Schematic diagram of the station for formation of lipid bilayer membranes. A. Polyethylene cup or beaker forming inner compartment. This container is thinned by heating in one area of the wall and punctured to form the supporting septum and aperture for the lipid membrane. B. External compartment. This may be of glass or transparent plastic. C. Aperture holding the membrane. D. Direction of light source for viewing the membrane. E. To the microscope for viewing the membrane. F. Electrodes for electrical observations. G. Enlargements of aperture region schematically showing the initial thick film formed by brushing the lipids across the aperture and the thin film formed by retraction of the excess lipid to the edges of the aperture.

found for cellular membranes. Subsequently, a number of different methods for membrane generation have been developed, using a variety of compartment configurations. Compositions have been varied widely to suit particular needs and to determine the conditions required for stable membranes.

These membranes have a number of desirable properties for experimental purposes. A primary advantage is the accessability of the two aqueous compartments, allowing easy insertion of electrodes, pipets, or tubes for electrical measurements or control of the ionic environment or for sampling the medium.

B. Composition

The composition of the membranes may be varied widely. Many of the early systems tended to have a complex composition, utilizing the natural mixtures of phospholipids obtained in tissue extracts as the primary amphipathic material. More recently, a wide variety of simple substances have been found to be effective in generating the lipid bilayers (17-19). Solutions containing only two components are commonly used now. Most commonly, membranes are formed with a phospholipid in a hydrocarbon solvent (hexane, decane, tetradecane), sometimes with a sterol added. Volatile solvents (chloroform, methanol, benzene, etc.) may also be included in the initial mixtures. Another commonly used mixture contains α-tocopherol as the plasticizing solvent in place of the hydrocarbon. But, in addition to the phospholipids, simpler molecules, long chain fatty acid mono- or diesters of glycerol (17, 18), sorbitan esters (17, 18) mono- or dialkyl phosphates or phosphites (19), quaternary amines (19), and other substances, may stabilize the lipid film.

It generally appears to be necessary to include two components, an amphipathic substance, which eventually forms the major portion of the membrane, and a solvent. The solvent itself may have some amphipathic properties, as in the case of tocopherol, which probably constitutes a significant portion of the molecules forming the bilayer when it is used. While it seems logical that some amphipathic molecules should be capable of acting as both surfactant and solvent, thus providing a single-component membrane generating system, no such substance appears to have been found so far. In general, a certain degree of heterogeneity in the lipid components appears to assist in maintaining lipid bilayer stability. This may be analogous to the cellular membrane which appears to maintain a sufficient heterogeneity to be able to adapt to diverse environmental conditions (20).

C. Structure

The nature of the material required for forming the black films, and their extreme thinness, immediately suggests that they any correspond to the bimolecular leaflet, or lipid bilayer structure proposed by Davson and Danielli (21), as the core of the cellular membrane. This assumption tends to be confirmed by the numerous measurements of thickness based upon optical (21-25), capacitive (15, 26-28), and electron microscope (15,16,29) studies. These all indicate an average thickness of about 70 Å, corresponding, approximately, to twice the length of the amphipathic molecules constituting them.

Measurements of interfacial tension in several laboratories (30-32) showing very low values of 0.1-6 dyn/cm also indicate that the surface of the membrane must be essentially polar, corresponding to the bilayer orientation.

III. LIPID BILAYER MEMBRANES: ELECTRICAL PROPERTIES AND ION TRANSPORT

A. Resistance

Despite the variety of substances which may be used in forming the membranes. the electrical characteristics tend to be consistently unspectacular, with no real hint of the strongly variable conductance evident in excitable cells. The resistance tends to be rather high, usually ranging from 10^6 to 10^8 Ω cm^2, although both lower and higher resistances may be found in special cases. This uniformity might be expected, to some degree, from the structure of the bilayer, in which the center core of the membrane should be essentially entirely composed of hydrocarbon with the varying polar groups all found in the interface region. The central region of all membranes would then have nearly the same dielectric constant (about 2) with a similar thickness, providing the main insulating barrier region. In spite of this, a number of properties of the basic lipid bilayer membrane are of interest in evaluating ion interactions in the excitable membrane.

Although the membranes seem to show nearly ohmic behavior at low potentials, there is generally a distinct deviation, toward a lower resistance, at higher polarizing potentials. Careful measurements, however, suggest that the deviation occurs at even very low potentials, as shown in Fig. 2, which was developed from numerous points obtained from X-Y recorder presentation of current-potential data, under potential control conditions.

B. Ion Permeability

The unmodified lipid bilayer membrane generally has an extremely low permeability to inorganic ions and measurement of permeability is

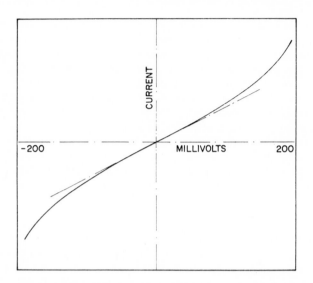

Fig. 2. Steady state current-potential relations of the basic lipid
bilayer membrane. Measurements were made on a membrane formed from
sphingomyelin, 5% and α-tocopherol, 40 %, in 2 : 1 chloroform-methanol
on a polyethylene septum. Current was recorded via the recorder output
of a sensitive ammeter (to 3 x 10-12 A full scale) directly as the Y-axis on
an X-Y recorder. Potential was similarly recorded as the X-axis through
a high input impedance (> 10^{14} Ω) electrometer. Recordings were made as
individual points, allowing equilibrium to be established for each potential
setting. Constant potential conditions (10^4 Ω series resistance) were used.
The curves show a consistent, nonlinear current response to polarization
although deviations are very small in the low potential regions.

complicated by the small membrane area generally used. However, it has
become apparent that the nature of ion permeation varies significantly in
different membrane systems. In the type of membrane used in this laboratory, containing tocopherol as a plasticizing solvent as well as a partic-
ipating surfactant, there is substantial evidence that only hydrogen or
hydroxyl ions carry significant currents across the membrane. Other ions,
with the possible exception of certain weak ions or complex forming ions,
appear to be almost totally excluded. This is indicated, first, by the lack
of significant diffusion potentials in either ion concentration gradients or in
bi-ionic gradients across the membrane (Table 1) while pH gradients produce large diffusion potentials (33). Although one should not expect a diffusion potential in an ion gradient if the transfer numbers of the ions concerned are similar, it is difficult to assume that the transfer numbers of
all the ion combinations tested should satisfy this criterion. It must be

concluded, then, that the inorganic ion permeabilities in these membranes
are so low that they cannot significantly affect the ion flow contributed by
hydrogen or hydroxyl ions at the concentrations at which these ions normally
exist due to dissociation of water. This conclusion is supported further by
the failure to show a change in membrane conductance upon varying the
NaCl electrolyte concentration from 0.1 mM to 0.1 M, as indicated in
Table 1. Noguchi and Koga (34) have obtained similar results with a number
of different membrane formulations, including those with a hydrocarbon
solvent replacing the tocopherol of the membranes used here, and they also
concluded that protonic conduction was the main transport mechanism in the
bilayer membrane. Their studies showed that nearly theoretical Nernst
potentials of 60 mV per pH unit gradient were obtained within the pH range
4 to 6.

On the other hand, a number of other investigators have obtained
opposing responses in various membrane systems. In some cases, large,
relatively stable membrane potentials are developed in response to ionic
gradients across the membranes (35, 36). Calculations based on those
potentials indicate that cation transfer numbers are independent of the
nature or charge of the phospholipid component (36). However, more recent
studies have tended to show a direct relationship of ion permeability to
charge on the membrane surface in certain types of membranes (73, 74).
In addition, the conductance of some membranes appears to be linearly
related to the salt concentration (19, 36), while in others, the conductance
may be constant at low concentrations with an abrupt break toward increasing conductance at higher concentrations (0.1-0.5 M) of electrolyte (19, 34).
Such transitions may suggest that increasing electrolyte concentration
exerts some effect upon the structure of the lipid bilayer or water-membrane
interface to cause a change in the energy barrier to the normal transport
ions.

A more direct observation of the effects of electrolyte concentration on
bilayer structure has been made here, using membranes formed with uncharged amphipathic molecules. Membranes generated from glyceryl
monooleate in decane or "sorbitan monooleate" (a fraction from Span 80 or
Span 85, Atlas Chemical Industries) in nonane are quite sensitive to electrolyte concentration and are stable only in relatively low concentrations of
sodium chloride. These membranes have long lives in low salt media but
may rupture soon after forming the bilayer in the higher (0.1-1.0 M) concentrations. This is but one of the structural influences which may be
attributable to the ionic composition of the aqueous phase.

It is difficult to asses the real meaning of the membrane potentials
used to indicate relative permeabilities in absence of actual measurements
of ion transfer across the membrane. Reliable results in such measurements

TABLE 1

Ion Permeabilities of Lipid Bilayer Membranes[a]

A. Ion concentration gradient potentials and bi-ionic gradient potentials

Reference electrolyte	Test electrolyte	Observed potentials (mV ± S. D.)	Cation transfer number
0. 1 M NaCl	0. 1 M NaCl	0. 0 ± 2. 2	
0. 02 M NaCl	0. 1 M NaCl	-0. 6 ± 0. 9	0. 50
0. 05 M NaCl	0. 5 M NaCl	-0. 9 ± 1. 2	0. 50
0. 01 M KCl	0. 1 M KCl	-1. 5 ± 0. 9	0. 50
0. 1 M NaCl	0. 1 M KCl	-1. 2 ± 0. 6	
0. 1 M $MgCl_2$	0. 1 M NaCl	-3. 2 ± 1. 6	
0. 1 M $CaCl_2$	0. 1 M NaCl	-4. 1	
0. 1 M $MgCl_2$	0. 1 M KCl	-2. 8	
5 mM Histidine, pH 6	5 mM Histidine, pH 7	+16	
5 mM Histidine, pH 6	5 mM Histidine, pH 8	+53	

B. Relation of membrane resistance to electrolyte concentrations

Electrolyte	Specific resistance (Ω cm^2 x 10^{-6})
0. 0001 M NaCl pH 7	18 ± 9
0. 001 M NaCl pH 7	20 ± 5
0. 01 M NaCl pH 7	13 ± 6
0. 1 M NaCl pH 7	9. 7 ± 3
1. 0 M NaCl pH 7	11 ±5
0. 1 M NaCl pH 6	12
0. 1 m NaCl pH 8	2. 9

have generally been difficult to obtain due to the high resistance and small membrane area. Pagano and Thompson (37), using tracer ions with spherical bilayer membranes, which have relatively large diffusion areas, have reported a sodium flux of 0.39 moles/cm^2/sec and a chloride flux of 90.2 moles/cm^2/sec in membranes containing egg phosphatidyl choline and tetradecane, with a chloroform-methanol solvent. The chloride value is much greater than would be predicted from electrical parameters and a carrier-mediated exchange mechanism was proposed to account for the discrepancy.

Petkau (38) has measured sodium tracer flux across flat lipid bilayer membranes generated in 0.1 M NaCl, water, or 0.005 M MgCl$_2$ and found limiting permeability coefficients of 3.71 x 10^{-8}, 0.27 x 10^{-8}, and 0.05 x 10^{-8} cm sec^{-1} respectively, in the three cases. These differences were attributed to solvent drag and the influence of ion density and potential profile at the membrane interface. The largest value reported here for a decane-phospholipid-chlolesterol membrane is roughly comparable with the value reported by Henn and Thompson for the spherical membranes of similar composition. Both values must be larger, by at least one order of magnitude, than the sodium permeability in phospholipid-cholesterol-tocopherol membranes where sodium transport was undetectable under conditions capable of detecting a flux corresponding to a diffusional permeability coefficient of 10^{-9} to 10^{-8} cm sec^{-1} (Bean and Shepherd, unpublished results).

Footnote to TABLE 1

[a]Membranes were formed from solutions of brain phospholipid, 1.5-2 %, α-tocopherol, 15-20 %, and cholesterol in 2:1 chloroform-methanol, across a 1.3-mm^2 area aperture in a polyethylene septum. For gradient measurements, the membrane was spread with both compartments containing the lowest electrolyte concentration.

The second concentration, or another electrolyte for bi-ionic gradients, was then obtained either by replacement of the medium of one compartment with an infusion pump arrangement or by adding an appropriate amount of a more concentrated solution. In part (B), all membranes were generated at pH 7 and then adjusted to the desired pH by adding acid or base to avoid the possibility of altering membrane composition and permeability, due to the effects of pH on the generating process.

Potentials and currents were measured through agar-salt bridges to calomel electrodes, with electrolyte concentration in the bridge similar to that in the test compartment to avoid a significant change in the test concentration due to leakage from the bridge or electrode.

Measurement of the change in membrane conductance with temperature, in phospholipid-tocopherol membranes, showed a logarithmic increase in conductance with increasing temperature in the range of $10°$ to $60°C$. There was generally some indication of deviation at lower temperatures. Arrhenius plots (log conductance versus $1/T$) gave straight lines with slopes indicating a 14-16 kcal/mole activation energy (see Fig. 3), even in different membranes which varied by as much as two orders of magnitude in resistance (39). These values are similar to those obtained for sodium flux in lipid liquid crystals (40). They are also not far from the activation energy of 14.6 kcal/mole for water transport through the lipid bilayers as determined by Redwood and Haydon (41).

It appears, then, that current carrying ions may vary in different types of bilayer membranes, with some substantial effects of the ions, themselves, upon the membrane conductance or stability.

C. Effects of Specific Ions on Ion Transport and Other Membrane Properties

Studies on the effects of various ions on the formation and conductance of tocopherol-phospholipid-cholesterol bilayers at pH 7 (33) seemed to indicate that there was little difference in thinning rate, stability, or membrane resistance in phospholipid-tocopherol-cholesterol membranes, attributable to the cations, sodium, potassium, magnesium, or calcium (Table 2). On the other hand, the anions seemed to influence both thinning rate and final resistances in a parallel fashion. Tenfold differences in either resistance or thinning rate appear with varying anions. It must be emphasized that the differences in membrane conductances in these cases can not be due to differences in permeability of the electrolyte ions, since none of the ions will produce a diffusion potential in a gradient or create a concentration dependent change in conductance. Instead, the changes must be attributed to the ionic influence upon the interfaces, leading to altered potential barriers for the normal current carrying ions, hydrogen, or hydroxyl.

In contrast with our results, which show little difference between membranes generated in electrolytes containing monovalent or divalent cations, Ohki and Goldup (42), working with phospholipid-hydrocarbon membranes, found that membrane resistances were markedly higher with calcium than with sodium at pH > 7. But the calcium and sodium curves cross at about pH 6 and are reasonably close at lower pH values. Thus, it appears that the cation can also specifically affect the membrane conductance, probably by an influence on the structure of the interface.

There are cases, however, in which the membrane becomes significantly permeable to inorganic ions. Ferric ion causes as much as a

1000-fold increase in conductance at concentrations less than 10^{-5} M in egg phosphatidyl choline-tetradecane membranes (36). This increase is associated with an apparent anionic permselectivity (43). The weak cations, aluminum and zinc, which may, like ferric ion, form complex ions, create large "diffusion" potentials, but cause little or no change in membrane conductance (33). Iodide also lowers the resistance of the lipid membranes drastically (44). This anion also causes an apparent anionic selectivity. The marked effects of ferric and iodide ions may be due to a carrier effect, in contrast with the structural effects of the other ions above, since both form complex ions with significant solubilities in organic solvents. However, suggestions have also been made that the redox characteristics of these ions may be involved in a semiconductor type of transport (44, 45).

As noted briefly above, some divalent or multivalent ions produce significant membrane potentials when distributed unevenly across the membrane. Since the most active members of this group (33) are weak ions which will form complex ions, it is possible that these represent true diffusion potentials, due to movement of the complex anion form. However, both magnesium ion and calcium ion also produce small membrane potentials when added on one side of the membrane as a minor component (1-25 mM) in a sodium chloride medium (0.1 M). These ions are not likely to diffuse through the membrane any better than the sodium ion and it seems probable that these potentials may be due to an interaction phenomenon related to those described below which alters the Donnan distribution of the interface.

An interesting effect of specific ions on the membrane has been noted by Papahadjopoulos and Ohki (46). Although a bilayer membrane may be quite stable when generated in a calcium medium, as described above, these authors found that the addition of low concentrations of calcium ion to one side of a phosphatidyl serine-decane membrane could cause loss of membrane stability. At concentrations below the rupture point, the small amounts of calcium caused a decrease in resistance of the membrane when added on one side only, even though the same concentration, added bilaterally, would generally increase the membrane resistance. This effect was dependent upon the pH, with maximum effect at higher pH values. With phosphatidylcholine membranes, the asymmetric addition of calcium had little effect on stability, indicating that the conductance and stability changes with the phosphatidyl serine must be due to specific interaction with the acidic sites of the interface.

D. Organic Ion Permeability and Interaction

With many organic cations and anions, a membrane potential develops when the ion is asymmetrically distributed across the membrane (47). This phenomenon is particularly marked with the analgesic amines, procaine, butacaine, and tetracaine. As shown in Fig. 4, the potential closely follows a Nernst diffusion potential relationship in its variation with concentration difference across the membrane. Similar potentials are found with tryptamine, tetracyclines, quinidine, and a number of other amines. Measurement of tryptamine flux across the lipid bilayer membrane by fluorescent assay proceudres (47) has shown that it diffuses through the

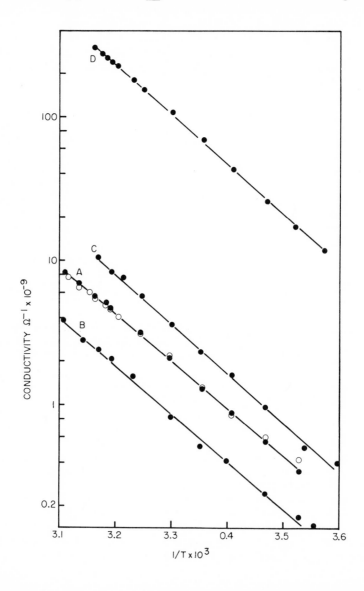

TABLE 2

Influence of the Electrolyte on Membrane Thinning
Kinetics and Membrane Resistance

Electrolyte (0.1 M)	Thinning time (sec)	Resistance (Ω cm^2 x 10^{-6})
NaCl	15 - 180	5
KCl	25 - 140	4.5
MgCl$_2$	30 - 180	5
CaCl$_2$	20 - 210	6.5
Na$_2$SO$_4$	200 - 600	16
Sodium phosphate, pH 7	360 - 1000	30
Sodium acetate, pH 7	15 - 150	3
Sodium lactate, pH 7	>1000	50

[a]Membranes were generated from brain lipids, 2 %; α-tocopherol, 20 %; and cholesterol, 2.5 %, in 2:1 chloroform-methanol on a polyethylene septum at 37°C. Thinning time is given from the time of brushing to the spreading of the black regions generally to the edges of the aperture, without consideration of lenses trapped in the black regions.

Fig. 3. Temperature-conductance relations in lipid bilayer membranes. The curves show reciprocal temperature versus log of conductance (Arrhenius plots) for several different membranes, varying more than 100-fold in conductance at any given temperature. Despite the variation in conductance between different membranes, the slopes of the curves are nearly identical, suggesting that the conduction mechanism is similar in all cases and the difference probably arises in the density of conducting sites. All membranes were formed on the same aperture at about 35°C. Temperature was monitored in the inner compartment with a glass-enclosed, chromel-alumel thermocouple. Current was monitored as indicated in Fig. 1. Current and temperature were recorded directly on an X-Y recorder. Each curve represents at least two traverses of the observational range, indicating a high degree of stability over the time period involved, A. Phosphatidylethanolamine-cholesterol-tocopherol membrane. B. Brain phospholipid-cholesterol-tocopherol membrane. C. Sphingomyelin-tocopherol membrane. D. Sphingomyelin-tocopherol (aged preparation) membrane.

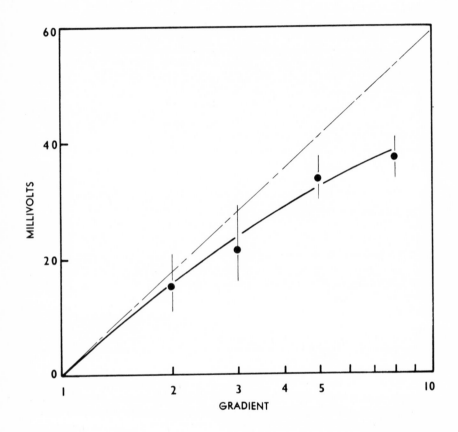

Fig. 4. Apparent diffusion potential relations with an asymmetric distribution of procaine across the lipid bilayer membrane. Potentials measured across a brain phospholipid-cholesterol-tocopherol membrane generated in 0.1 M NaCl, pH 7, after adding procaine solutions, adjusted to pH 7, to each side to establish the indicated concentration gradients. Reference side contained 1 mM procaine, with higher concentrations on the experimental side of the membrane, giving rise to negative potentials, consistent with cationic diffusion. Butacaine and tetracaine produced similar potentials.

membranes too rapidly in comparision with the neutral analog, indole-3-ethanol, to account for the permeability only by diffusion of the neutral form of the amine at the concentrations which could be expected in the bulk aqueous phase. Instead, the diffusion potentials and permeability seem to indicate that these physiologically active amines may be transported through the membrane in a cationic form. Alternatively, one must assume an alteration of the dissociation equilibrium in the interface in such a manner as to provide a higher gradient of the ion than is apparent from bulk phase equilibria. The latter mechanism is unlikely thermodynamically, although it might operate in the special situation of near zero concentration on one side of the membrane.

Short chain aliphatic amines do not produce the same type of diffusion potential phenomenon as the complex, physiologically active amines. However, long chain alkyl amines and quaternary ammonium ions will produce a membrane potential [(48), and R. C. Bean, unpublished data]. Since it has been demonstrated that some of these amines may stabilize a bilayer (19), it seems possible that these membrane potentials arise from an asymmetric adsorption or insertion, again creating an altered Donnan distribution. There has been no measurement so far of actual flux of the long aliphatic amines through the membrane to ascertain whether permeability is sufficient to create a true diffusion potential rather than an interface potential. Another possibility that must be considered arises from the effect that some cationic surfactants have on membrane conductance. Seufert (49), showed that low concentrations of the surfactants may greatly enhance the conductance of the lipid bilayer membrane. Thus, the potentials could be related to an asymmetric diffusion region in the membrane, produced by the amines.

Many organic anions also increase the conductance of the membrane greatly (50-52). Some of the most effective inducers are also the most effective uncouplers of oxidative phosphorylation. The change in conductance, in these cases, can be entirely attributed to an increased permeability for hydrogen (or hydroxyl) ions (51), suggesting that the anions are acting as proton carriers.

A number of other organic substances, such as dipicrylamine and tetraphenylboron can also increase the conductance of the membrane, apparently by acting as cation complexors and carriers (16, 53).

Several cationic polyelectrolytes, protamine, acetylcholinesterase, polyornithine, polyarginine, and cytochrome C, for example, may either increase membrane conductance or destabilize the lipid bilayer [(16), R. C. Bean, unpublished data] in a manner related to the effects of the divalent ions noted above. In contrast, anionic polyelectrolytes, like ribonucleic acid, appear to have little effect.

It is apparent that lipid bilayer membrane properties may be significantly altered by the interactions with numerous cationic substances, both inorganic and organic. Some of these effects may be recalled in discussing the properties of the excitable membrane.

IV. THE EXCITABLE LIPID BILAYER MEMBRANE: GENERAL DESCRIPTION

A. Development of Excitability

If a small amount of a protein substance, designated EIM (for "excitability inducing material") by its discoverers (12, 13), is added to the electrolyte solution on one side of the lipid membrane, the conductance of the membrane at low polarizing potentials increases rapidly, as shown in Fig. 5. Conductance may change, in some cases, by as much as five orders of magnitude, lowering the resistance to values in the range of 100 to 1000 Ω cm^2. If the current-potential characteristics of these membranes are now examined under potential control conditions, I(V) curves similar to those of Fig. 6 may be found. This shows a region of high conductance (the open state) at low polarizing potentials, which switches at higher polarizations through a negative resistance region to a much lower conductance (closed state). A closed state develops with either positive or negative polarizing potentials. (Note: in the discussions here, a positive potential or current will indicate a cation current toward or into the compartment containing EIM; a negative current, a cation flow away from the EIM compartment.)

Current responses to potential control pulses in a membrane of this sort are illustrated in Fig. 7(a), comparing positive and negative pulses (taken from oscilloscope recordings). Fig. 7(b) shows the potential responses to constant current pulses. The transition to high resistance develops with a characteristic time constant at a threshold potential, creating in abrupt potential increase. Membrane I(V) relations are characteristically asymmetric. The closed state transition is usually attained at lower polarizing potentials of positive polarity than with the negative polarity. In some special cases, the curves may become nearly symmetric or even reversed. Symmetry may develop if EIM is added to the membrane on both sides rather than one, but switching to the closed states then generally develops on both sides at potentials equivalent to the higher of the two switching regions of the normal, asymmetric membrane. Since the potential required for the negative switching may be above 100 mV in some lipid systems, the symmetric membrane may appear to have an almost linear resistance within most of its dielectric stability range (about 100-150 mV).

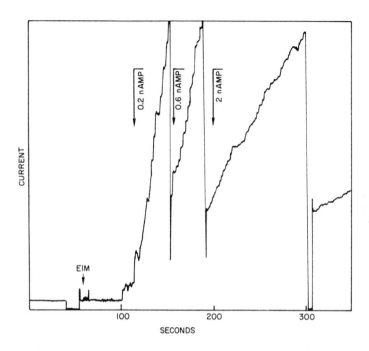

Fig. 5. Kinetics of the lipid bilayer membrane conductance change during interaction with EIM. The phosphatidylethanolamine-cholesterol-tocopherol membrane was generated in 0.1 M NaCl, pH 7, and allowed to attain a steady resistance value. While polarizing at - 10 mV, EIM was added at the point indicated. Following a short induction period (which varies with membrane composition, EIM concentration, temperature, and depth or width of the aperture), current at constant - 10 mV polarization increases rapidly. The curves represent tracings of direct recordings of the current with time. As current increased with changing conductance, ammeter output was attenuated as indicated. The current shows increasingly large fluctuations with increasing extent of reaction. Attenuation at the higher current values masks the continued increase in low frequency noise. Stirring was continuous throughout the experiment.

B. Nature of the EIM

The EIM is a protein-containing material, best derived from the growth of Enterobacter cloacae, ATCC 961 (formerly Aerobacter cloacae). It is produced by the bacterium in defined glucose-salt media, largely in the log growth phase of the bacterium. Growth conditions may be defined so that the active material is produced into the growth medium (54) or retained in the bacteria and extracted by high pH extraction (55). Kushnir's studies

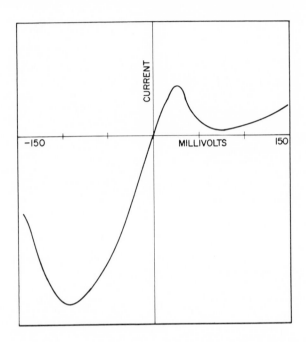

Fig. 6. Basic, steady state, current-potential relations for the EIM lipid bilayer membrane. The curve represents an average steady state relation for a brain phospholipid-cholesterol-tocopherol membrane and indicates the existence of two negative resistance regions and a high degree of polarization asymmetry.

(54) indicate that the EIM may consist of two components, a protein fraction and a ribonucleic acid fraction, neither active alone but highly active in combination. The highest activity preparations will cause significant changes in membrane conductance at concentrations below 1 ng/mliter, and even these preparations may be composed largely of inactive impurities. The EIM elutes from Biogel 150 gel permeation columns (exclusion limit 150, 000 m. w.) at or near the elution front indicating a relatively large molecular weight, although it may be a smaller molecule associated with larger molecular impurities in the preparations. Assuming a molecular weight of 100, 000, the EIM would be active at concentrations less than 10^{-12} M, or somewhat more effective than the highly active cyclic depsipeptide, valinomycin, in promoting ion permeability. In the impure form it is not readily heat denatured, but its activity disappears rapidly with proteolysis. Even after the EIM has reacted with the membrane to lower the membrane resistance and produce the negative resistance, the resistance may be raised once more through proteolytic action nearly to the original, bare membrane

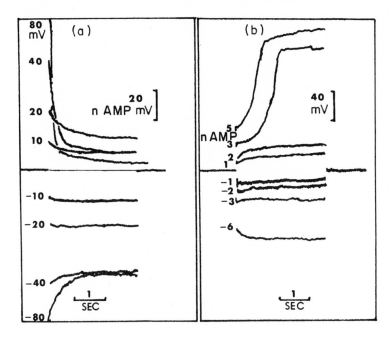

Fig. 7. Kinetics of the response of the EIM membrane to pulse stimuli.
(a) Current response to potential control pulses. Tracings of oscilloscopic
recordings of the change in current upon stimulation with pulses of the indicated potentials. Positive pulse sequences were completed before the
negative pulse sequence, increasing potentials sequentially, with adequate
rest in between each pulse to allow full recovery to the resting conductance.
(b) Potential response to current control pulses. Tracings of recordings
as above, using a high resistance in series with the membrane resistance
to maintain current control.

level (16, 55). In the crude, or partially purified form, EIM activity was
not affected by ribonuclease. However, Kushnir (54) states that the high
specific activity EIM fractions are inactivated by ribonuclease. This inactivation could still be a proteolytic action, as the result of greater sensitivity to the small amounts of protease usually present in ribonuclease
acting upon the very small amounts of protein in the purified EIM.

C. Mechanisms for the EIM Conduction

The EIM channel is normally almost completely specific for cation
transport. This is indicated by the nearly theoretical Nernst potentials
developed in ion concentration gradients (16, 33) and by tracer measurements
of ion transport, as demonstrated in Fig. 8. This figure compares, during

the development of the EIM conductance, the total ion flux, calculated from the current, with the tracer cation flux for ^{24}Na and ^{45}Ca. Prior to adding EIM to the membrane no tracer flux can be detected, even after several hours with high specific activity ions, for sodium, calcium, iodide-^{131}I, or phosphate-^{32}P. For the anions, even after the EIM conductance had reached its optimum level, no tracer flux could be detected, with conditions which would permit detection of the tracer at about one per cent of the total ion flux. But sodium and calcium tracers became readily detectable soon after the EIM conductance began to develop. In each 10-min sampling period, after the tracer cations reached a measurable level, the cation flux was able to account for the total ion current across the membrane. The quantitative relation was maintained at higher polarizations, which developed a partly closed state and reduced the ion flow. The quantitative relations are somewhat clouded, however, by the counter flow of sodium ions. With currents opposing the flow of the tracer ions, the calcium tracer flux was completely inhibited, but the sodium tracer flux was only reduced by about two-thirds. Thus, it appears that the EIM channel is essentially a single file channel so far as the calcium is concerned, while it permits relatively large counter flow exchange of the sodium ion. Both the tracer and concentration gradient experiments indicate that the EIM conduction is at least 98% selective for cations over the anions.

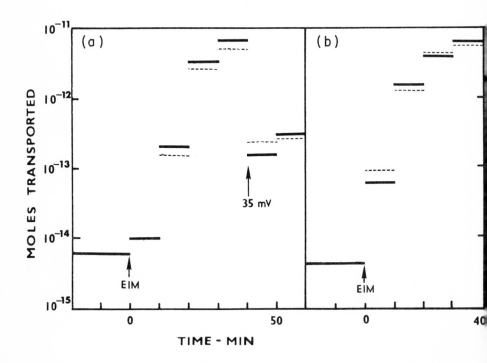

Bi-ionic potentials (33) indicate that there is only slight discrimination between different univalent cations. Recent direct measurements of conductance in membranes with single EIM channels show the following relative order of cation mobilities in EIM channels (NaCl = 1.0): Li, 0.73; Na, 1.0; K, 1.5; Rb, 1.6; Cs, 1.3; NH_4, 1.4; $\frac{1}{2}Mg$, 0.7; $\frac{1}{2}Ca$, 0.7 (75).

Bean et al. (55) have shown that the EIM conductance develops through a series of discrete current jumps of relatively uniform size, rather than through a continuous change in conductance. This phenomenon, illustrated in Fig. 9, suggests the formation of discrete unit channels by penetration or interaction of the EIM in the membrane. This tends to be confirmed by the observation that the conductance can be destroyed in the membrane even by endo-peptidase attack across the membrane from the side to which the EIM has been added (16, 55). In conjunction with the evidence that the EIM conductance is not diminished by washing out the excess EIM after the

Fig. 8. Tracer cation transport in lipid bilayer membranes during development of EIM conductance. (a) Flux of ^{24}Na ion. Solid bars, total ion flux; Dashed bars, ^{24}Na ion flux. Solid bars, total ion flux; Dashed bars, ^{24}Na ion flux. A sphingomyelin-tocopherol membrane was generated in 10 mM NaCl and then part of the solution of the inner compartment was replaced with $^{24}NaCl$ to give an activity of 2.47 x 10^7 cpm/μmole, with liquid scintillation counting. The base line transport and conductance was established, taking 0.2 ml samples for counting from the outside compartment (and replacing with fresh solution) at three 10 min. intervals, while recording current at a constant + 10 mV polarization. EIM was then added, maintaining the same constant polarization. Samples were taken for counting from the external compartment at 10 min. intervals while maintaining current recordings. Current was integrated over each 10 min. sample period for computing total ion flux while cation flux was calculated from the specific activity of the tracer ion and the measured transport activity. After 40 min. of EIM reaction, the polarizing potential was change to 35 mV to switch to a high resistance state, with consequent reduction in both cation transport and in current. It must be noted that even though there appears to be a close correspondence of tracer cation transport and current ion flux in this experiment, measurements made with polarization opposing tracer cation transport in other membranes indicated that tracer flux continued at about one-third the rate observed with a favorable polarization for tracer transport. (b) Flux of ^{45}Ca ion. Solid bars, total ion flux; Dashed bars, ^{45}Ca ion flux. Conditions as in (a), except utilizing 5 mM $CaCl_2$, with a specific activity of 1.13 x 10^7 cpm/μmole in the inside compartment. With calcium ion, a polarization opposing the tracer ion transport completely blocked movement of the tracer across the membrane within experimental error.

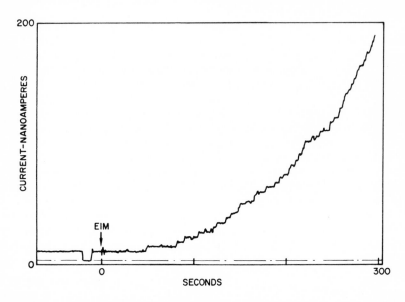

Fig. 9.   Discrete conductance changes during the reaction of EIM with the lipid bilayer membrane.   Conditions are essentially those of Fig. 5, except that a lower concentration or EIM was used to reduce the rate of reaction with the membrane.   Shortly after introduction of the EIM, a step change in current develops, followed by further step increases at an increasing frequency.   More than fifteen individual steps, each corresponding to a conductance change of $4.0 \times 10^{-10}\Omega^{-1}$,   may be observed with several more corresponding to 8.0 or $12.0 \times 10^{-10}\Omega^{-1}$.   Two smaller steps may be noted.   Temperature was $36°C$.

conductance has been established, this means that the peptidase reversal is probably not due to destruction of a carrier or alteration of an equilibrium between aqueous and lipid phase.   It implies, therefore, that the EIM must form a protein bridge across the membrane providing a polar, ion conducting channel, with a portion of the protein accessible to attack of the peptidase on the other side of the membrane.   Such attack must destroy a configurational requirement for the open channel.   Further support for this concept is indicated from a careful observation of the kinetics of the proteolytic reversal (55) in which conductance steps similar to but smaller than those found during conductance generation, may be found.   After proteolytic reversal has been completed and the two-state behavior completely eliminated, there is generally a small, residual conductance above that of the original, basic lipid bilayer membrane, as if some residue of the protein remained in the membrane providing a very small conducting path.

These experiments, together with confirming experiments by Ehrenstein, Lecar, and Nossal (56), indicate an open channel conductance of about 3 to $4 \times 10^{-10} \Omega^{-1}$ per channel, varying from a low extreme of 2×10^{-10} to a high of $6 \times 10^{-10} \Omega^{-1}$. If this value is assumed to be that of the unit conducting element, it becomes much too great for a carrier mechanism, requiring a flow of about 10^8 ions per sec via each carrier unit. This would be much too large a turnover to be accommodated by a carrier with a molecular weight of 10,000-100,000 (57).

There appears to be no relation of this conductance value to the lipid composition of the membrane. In a large number of formulations (55) the variation in conductance jump was about as great between membranes of similar composition as it was between membranes of varying composition. Thus, it appears that the open state conductance of the EIM channel is essentially a function of the EIM alone, and cations being transported through the open configuration are essentially shielded from the influences of varying polar groups of the lipids.

Switching between open and closed states was also shown by Bean et al. (55) to develop through stepwise rather than continuous changes in current (Fig. 10). The steps,in membranes with only a few channels (less than twenty), tended to be smaller than those observed in channel formation, sometimes only amounting to one-fifth the normal formation step. This led to a proposal that the EIM channel may have several intermediate quasi-stable states between the fully open and fully closed states. However, Ehrenstein, working with membranes having one to five channels (56), concluded that there were no intermediate levels and the channels existed only in two states, open or closed. The smaller steps observed by Bean et al. may have been due to additive effects of the more numerous channels or to secondary effects found in some membranes discussed below.

Ehrenstein's measurements in an oxidized cholesterol-decane membrane indicate that the closed channel conductance was about $0.5 \times 10^{-10} \Omega^{-1}$ or about one-seventh of the open state conductance. This would provide essentially only a six or seven to one ratio for the open to closed state conductance. This is in conformance with characteristics of oxidized cholesterol membranes with large numbers of channels, in which the open state conductance slope is generally only five to ten times the closed state slope. On the other hand, a number of phospholipid-tocopherol membranes show open-to-closed slope ratios of greater than 100 (Fig. 11) or even 1000 (16), suggesting that closing can be much more efficient than that of the single steps obtained in oxidized cholesterol membranes. This was subsequently confirmed in studies on membranes of varying lipid composition having only one to three EIM channels. It was possible to demonstrate under these circumstances that the EIM channel generally has at least three

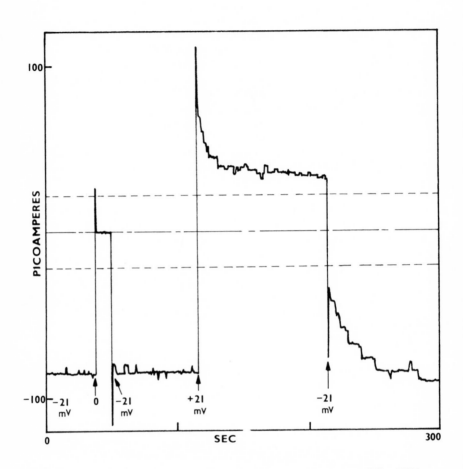

Fig. 10.   Discrete conductance steps during transitions of the EIM membrane between high conductance and low conductance states.   Membranes were prepared and reacted with a minimum amount of EIM, as in Fig. 9, but maintaining a lower temperature (25°) to minimize the rate of EIM reaction.   After development of 8 channels ($30 \times 10^{-10}\Omega^{-1}$) the polarizing potential was switched from maximum conductance at - 20 mV, to about 80% conductance at + 20 mV.   A partial conductance change occurs during the sweep time of the recorder used, but it is possible to distinguish several large conductance steps and a number of smaller steps following the transition, indicating that the switching transition also develops by the transition of individual channels between open and closed states.   The relations between the larger and smaller steps are discussed in the text.

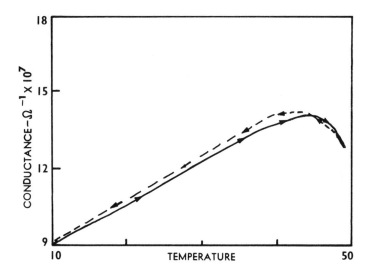

Fig. 11. Temperature coefficient of conductance in the EIM lipid bilayer membrane. The sphingomyelin-tocopherol membrane was generated in 0.1 M NaCl, pH 7, at 35° and allowed to react with EIM at that temperature until the reaction was essentially complete. The membrane bath was then cooled to 9° and the recording of temperature vs conductance started. Temperature was raised at about 1° per min, recording points every 15 sec. After reaching 47°, the cooling cycle was initiated, with a similar rate of temperature programming. The hysteresis evident between the cooling and heating curves is due partially to continued slow reaction with EIM at the elevated temperature, and partly to small thermal gradients which develop across the membrane due to the poor thermal transfer between internal and external compartments making the temperature record lag slightly behind the actual membrane temperature.

conductance levels or states, with the lowest state always having less than 2% of the open state conductance and a second state normally having a conductance about one-fifth that of the open state, corresponding to the lowest state found in the oxidized cholesterol membrane Ehrenstein et al. Still another intermediate level, corresponding to about one-half the open state, is also frequently found.

Temperature studies on the EIM membrane (39) show a transition from a relatively low, uniform positive coefficient for conductance at low temperatures to a steep negative coefficient at the higher temperatures as illustrated in Fig. 11. This anomalous behavior is somewhat clarified by an examination of the change in I(V) characteristics in similar regions as in Fig. 12. At the lower temperatures, the slope of the central portion of the curves (open state) increases with increasing temperature, while both the negative resistance regions move inward to lower potentials. This means that the potential required to develop the closed state continuously decreases. At the elevated temperatures the slope of the central portion of the curve representing the open state decreases, and the negative resistance portion of the positive leg may be entirely lost with a rectifying characteristic substituted. The I(V) relations have been replotted below as a function of per cent of open state vs potential for each temperature. In making these conversions it was assumed that the true open state conductance at all temperatures should be reflected in a continuous rise of the

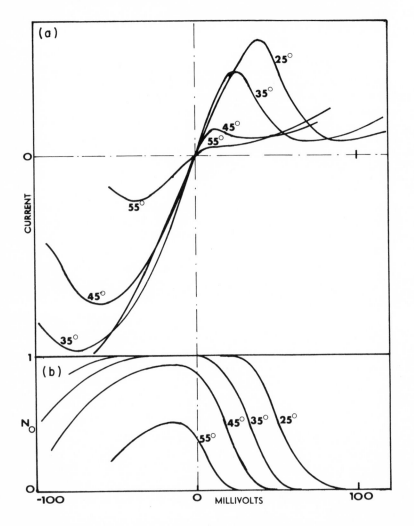

linear slope of the low temperature portion of the curve in Fig. 11. The
extrapolated, high potential conductance at each temperature was used inso-
far as possible as the residual closed state conductance to be subtracted
from all conductance measurements to provide a zero baseline.

The resulting parallel S-shaped curves for the positive end of the I(V)
relations are consistent with an equilibrium, following Boltzmann statistics,
between two stable states (open or closed) having a free energy difference
which may be affected by the polarizing potential. This two-state model
has been treated theoretically by Mueller and Rudin (14). The Mueller and
Rudin treatment for two states may be applied equally to either the positive
or negative legs of the EIM system at low temperatures, where the equilibria for the positive and negative transitions are essentially independent.

Fig. 12. Change in current-potential relations of the EIM membrane
with temperature variation. (a) Variation in current vs potential with
temperature. A sphingomyelin-tocopherol membrane was formed and
reacted with EIM at $45°$ to obtain rapid reaction to an end-point. Current
potential curves were then developed at $25°$, $35°$, $45°$, and $55°$, with repe-
titions at the intermediate temperatures before and after cooling to the
$25°$ level. (b) Distrubution of channels between open and closed state. The
fraction of channels remaining in the open state at any potential has been
plotted from the relation,

$$N_O = \frac{n_O}{n_T} \cong \frac{G_v - G_c}{G_{max} - G_c}$$

where N_O is the fraction of channels in the open state, n_O is the number of
open channels, n_T is the total number of channels, G_v is the observed conductance at a potential, V, G_{max} is the maximum conductance, corresponding to 100% open channels, and G_c is the conductance of the fully closed
state at positive polarization. The maximum conductance for the higher
temperatures was calculated by extrapolating the linear increase in maximum conductance observed between $25°$ and $35°$ on the assumption that the
conductance of individual open channels should continue to increase linearly
over the temperature ranges concerned (with some justification based upon
the change in the individual conductance steps observed during EIM insertion into the membranes at varying temperatures) and the decrease in
maximum conductance at the higher temperatures is due to failure to reach
100% open state. The positive closed state conductance value for each
temperature was used for both positive and negative legs of the curves due
to the difficulty in measuring the closed state conductance on the negative
end.

But at higher temperatures it is apparent that the two equilibria overlap,
reaching a point where all channels cannot be forced into open state because, as the potential is shifted in one direction or the other from the
central region, the equilibrium is simply shifted more in favor of one or the
other of the closed states. In extreme cases, which have been observed
with special lipid formulations (39), an increase in temperature may actually force all channels into closed states just as effectively as raising the
potential. Aside from its importance in possible interpretation of anomalous temperature coefficients in some biological phenomena (58), this information suggests that the EIM membrane must be treated as a three-state
system, consisting of three stable states, an open state, and two closed
states--one negative and one positive, in unimolecular, linear equilibrium.
Thus, in an analysis of a membrane with only a few channels, such as those
investigated by Ehrenstein, it may be possible to distinguish between conductances of the two closed states if there is any asymmetry in their conductance properties. Although the data with membranes containing large
numbers of channels indicate that the ratio of open state to closed state
conductance is usually larger for the positive transition than for the negative transition, the difference has not been sought or confirmed in single
channel experiments. It must be emphasized, however, that such experimental distinction depends on utilizing a membrane in which, firstly,
residual single channel conductances are large enough to measure over
background noise and, secondly, that this residual conductance is significantly different in the two closed states. Such conditions may be difficult
to meet due to the low residual conductances in those membranes which
appear to have a significant difference in the two closed state conductance
slopes. Nevertheless, some consideration may be given to the possibility
that some of the variability in few channel observations (55, 56) may be
partly due to failure to make this distinction.

All these data, taken together, indicate that the conductance and the
negative resistance behavior of the EIM membrane is due to penetration of
the membrane by large numbers of a complex protein to form numerous ion
conductive channels which may exist in either open or closed states. The
rates of transition between the open and closed states follow normal statist-
ical reaction kinetics, but they are influenced by polarizing potentials leading to a population redistribution between the open and closed states as a
function of potential with commensurate change in the resistance of the
membrane. This mechanism is closely related to "two-stable-state" mechanisms which have been proposed by some investigators (59, 60) for the
activity of excitable cells. Several types of change in channel characteristics or configuration could be responsible for the transition, but none can be
unequivocally developed at this point for lack of information on the specific
groups involved.

D. The "Steady State" Properties of the EIM Membrane

It is rather difficult to define steady state characteristics of the EIM membrane for several reasons. First, unless the excess EIM in the medium is removed by replacement of the medium, there is a strong tendency for the EIM reaction to continue slowly, almost indefinitely. There are also several types of hysteresis in some kinds of membranes so that the characteristics measured depend rather markedly on the recent past history of the membrane. Some of these hysteresis effects are of some importance in describing the relation of membrane characteristics to lipid composition and in the effects of divalent cations so they are discussed here.

One type of hysteresis is indicated in the current responses to potential control pulses in Fig. 13a. Starting at zero polarizing potential, a 30 mV pulse creates a partial switching to the closed state, reaching an equilibrium "steady state" current rapidly. An additional pulse to 80 mV causes a transition to the steady state closed conductance. But, upon depolarizing to 30 mV once more, the 'steady state" current differs markedly from that obtained in the increasing polarization step. Upon resting once more at zero potential, the cycle may be repeated entirely with the same current values registered.

This behavior is characteristic, to a greater or lesser degree, of most EIM membrane systems. It is generally accentuated by oxidized or polymerized impurities in the lipids, heavy metal ions, and other, indeterminate factors that tend to rigidify the lipid matrix.

A second hysteresis depends upon the polarity of the polarizing potential, as illustrated in Fig. 13b. A sphingomyelin-tocopherol membrane was polarized at a negative potential to a steady state current, giving about the optimal, open state conductance. Upon pulsing to 25 mV, very little change in resistance develops. A further pulse to 70 mV causes a rapid transition to a closed state. After a short return to zero polarization, another 25 mV pulse creates an initial current peak almost as high as the current obtained at 25 mV in the first cycle, but with a rapid decay as the resistance increases almost to that of the fully closed state. A repetitive cycle between zero and the positive polarization may continue to repeat this latter low energy transition, but upon return to a negative polarization for a short time, the channels become resistant to closing at low potential.

The hysteresis in either of these hysteresis loops develops with a characteristic time-potential relation so that minimal or maximal hysteresis may be obtained through time or potential control. The second type of hysteresis provides an interesting type of I(V) response under conditions where a negative polarization is applied only long enough to allow part of the channels to shift to the high energy transition state. Under these

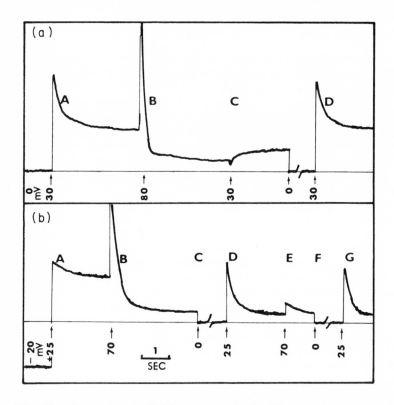

Fig. 13. Current-potential hysteresis in the EIM membrane. (a) Closed state to open state transition inhibition. The brain phospholipid-cholesterol-tocopherol membrane was allowed to rest at zero polarizing potential after the EIM reaction. A small positive potential pulse causes a partial shift into closed state (A). A higher potential pulse (B) causes a transition to the full closed state. A subsequent return to the low polarization (C) produces a lower "steady state" current than was observed for the original low polarization pulse. After a short rest at zero potential (D), the current at low polarization (E) is identical with the original pulse. Curves are tracings from original recordings. (b) Open state to closed state transition inhibition. The sphingomyelin-tocopherol membrane, after reaction with EIM was rested at - 20 mV polarization. A + 25 mV potential pulse creates only a small shift toward the closed state (A). A higher potential pulse (B) causes a rapid transition to a closed state. Subsequent cycles of polarization, starting with zero polarization show a rapid transition to closed state even at the low potential where the transition could not be observed originally (C to G). The peak heights on the initial current spikes indicate that there is little difference in the conductance at the beginning of cycle A to C and C to G. A negative polarization will cause a return to the original responses.

circumstances, the I(V) curves (obtained with a ramp potential; see below) may show a double negative resistance region, as those channels capable of undergoing the low energy transition first switch at the normal switching potentials and then; at higher potentials, the remaining open channels turn off via the high energy transition.

These relations are indicated in the modified, general "steady state" I(V) curves of Fig. 14. These show in 14(a), the hysteresis loop resulting from high potential polarization and in 14(b), the loops resulting from negative potential polarization. The limiting characteristics of the two state membranes, the open state conductance, G_O, and closed state conductance, G_C, remain the same in all cases. It is simply the route between these points that changes with consequent alteration in transition time constant. For convenience in analysis and discussion, the potentials corresponding to the maxima of the I(V) curves ($dI/dV = 0$) in both negative and positive quadrants have been used as a reference point and given the designation of "switching point" or "switching potential." At this point about three-quarters of the total channels originally in the open state at the potential of maximum conductance will remain open with one-quarter in the closed state. The three switching potentials are specifically designated as S1-, S1+, and S2+, for the negative and the positive low energy, and the positive high energy maxima, respectively. For purposes of thermodynamic calculations, the use of the potential at which open and closed states are equal ($N_O = N_C = 0.5$) might be more desirable. However, this half point must be derived from calculated data from the I(V) curves with certain assumptions that may not be entirely valid on the basis of other studies. Thus, for limited evaluations, such as those presented here, the easily observed peak potential seems adequate.

The curves shown here may be considered reasonably representative of those which will be obtained either from pulse analysis, giving a point by point description of the steady state values, or by ramp potential analysis (61, 62) which presents a continuous recording of the I(V) relations. In the latter case, I(V) curves may be recorded directly with an X-Y recorder or oscilloscope by changing the polarizing potential at a constant rate to the desired endpoint and then returning to the starting point with the same constant rate of change. While this process distorts the curves particularly in the region of the negative slope due to the transitional time constants and capacitive currents, it is generally more convenient than pulse analysis for defining the membrane parameters and determining their changes with experimental treatment. The ramp potential analysis has, therefore, been used in most of the subsequent analyses presented here.

One more contrary aspect of the "steady state" analysis may be brought to attention here. It may be noted in Fig. 7(a) and 13 that the resistance

transitions, particularly at high potentials, may develop with two distinct
time constants, one a fast reaction accounting for the major part of the
current decrease in switching, the other a slow change of much smaller
magnitude. This slow transition is even slower for switching on than for
switching off and appears to be a required preceding step in the on transition. In some instances, if the slow phase of switching off is allowed to
go to completion, a subsequent transition to the open state may be almost
completely inhibited. There is some suggestion that this secondary transition may be equivalent to partial reversal of the channel formation, perhaps
by an electrophoretic movement of the protein, so that the return to open
state also involves partial regeneration of the channel rather than a simple,
low energy site redistribution or configurational change.

E.   Effects of Lipid Composition on the EIM Membrane

As indicated previously, the lipid bilayer membrane may be formed
from a wide variety of lipids and solvents. EIM will not react with all types
of membranes, however. It reacts readily with many membranes containing

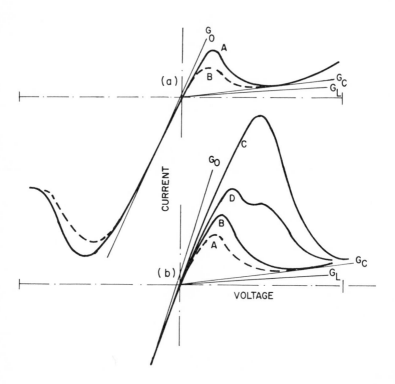

a phospholipid plus tocopherol or an ester for tocopherol. A hydrocarbon solvent with phospholipids inhibits the reaction, but an adequate excitable membrane may still develop if sufficient time is allowed. A number of synthetic lipids are also effective. However, the only formulation containing no ionizable molecules which is known to react well with the EIM is oxidized cholesterol (17), in hydrocarbon solvent. No reaction can be obtained with glycerol monooleate in decane or sorbitan oleate in decane (Bean, unpublished results).

Also, as previously indicated, the composition affects the membrane conductance parameters, the ratio of open to closed states. Additional effects are noted in the positions of the switching potentials, symmetry, and the nature or extent of hysteresis. A few representative I(V) curves for membranes of varying composition are given in Fig. 15 to indicate the nature of these variations. These show that the ratio of the two limiting conductances, G_O and G_C, may vary widely. In some cases, the hysteresis is almost absent while in others both types are major factors.

Transistion rates are also affected by the lipid composition and purity. Substitution of certain esters for the tocopherol in phospholipid-containing membranes may create a time constant of minutes for the conductance transitions rather than the fractions of a second found in the original, tocopherol-containing formulation (Bean, unpublished data).

Fig. 14. Generalized current-potential relations for the EIM lipid bilayer membrane. (a) General curves showing hysteresis due to closed state to open state transition inhibition. Curve A is produced by plotting points obtained with increasing polarization. Curve B is derived from decreasing polarization. G_O, maximum open state conductance; G_C, closed state conductance: G_L, conductance of basic lipid bilayer membrane; S1-, negative switching potential. S1+, positive, low energy (noninhibited) switching potential. (b) Hysteresis and double negative resistance regions due to open state to closed state transition inhibition. Curve A and Curve B represent the normal hysteresis loop obtained with little or no exposure to negative polarization, as in part A. Curve C develops during increasing positive polarization after a period of exposure to a negative polarizing potential. Curve D is a hybrid curve which may occur in an increasing positive polarization cycle after a short exposure to negative potentials so that the transition inhibition is only developed for some channels or in some membrane systems in which only part of the channels appear to be subject to the inhibition. After all channels have been switched to the closed state at high positive polarizations, subsequent cycles, not involving a negative potential, will follow the hysteresis cycle of A and B. S2+, high energy (inhibited) positive switching potential.

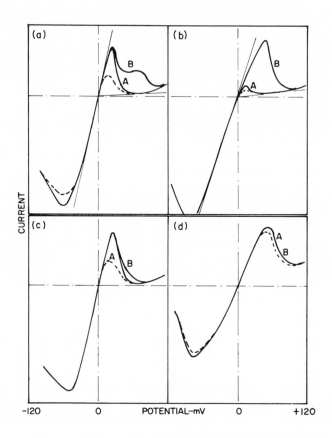

Fig. 15. Change in current potential relations with variation of lipid composition in the EIM lipid bilayer membrane.   Curves obtained by ramp potential measurements.   Curve A, in each case is traced after resting at zero or low positive polarizations.   Curve B is obtained after a negative polarization sufficient to obtain maximum inhibition of the open state to closed state transition, if this inhibition is possible.   Dashed curves indicate decreasing polarization traces.   (a) Phosphatidylethanolamine-cholesterol-tocopherol, 2:2:20.   (b) Sphingomyelin-tocopherol, 5:40. (c) Phosphatidylethanolamine-decane, 1:99.   (d) Oxidized cholesterol-decane, 2:98.

It is not surprising that the lipids exert a significant effect upon the EIM membrane parameters. In a channel formed by penetration of the EIM through the membrane, there must be ionic, hydrogen, and hydrophobic bonding or interaction between the EIM and the lipid lattice to stabilize the system. These interactions may be expected to vary with charge and polar group distributions in the membrane-water interface and the spacings and conformations of the molecules in the hydrocarbon core. They should exert a modifying influence upon any structural or conformational changes in the EIM involved in the conductance transitions. Therefore, even though the conducting channel may be entirely limited to the EIM, as indicated earlier, the lipids would have their effect upon the energy values of the transitions.

It is actually more surprising that gross variations in lipid, such as changing from a totally uncharged system, as in the case of oxidized cholesterol in decane, to one that is neutral but with a high density of fixed ionic sites, for sphingomyelin or phosphatidylcholine in tocopherol, or with a net negative charge, as for phosphatidylethanolamine, does not cause more drastic changes in membrane characteristics. This tends to confirm the basic assumption that the open channel of the EIM membrane is essentially contained within the protein complex and is not subject to direct control by the polar groups of the lipids. On the other hand, these interfacial alterations should have obvious effects in Donnan distributions and, consequently, on the energies required for movement of a cation through the interface and into and out of the EIM channel.

The lipid composition studies do little to clarify the nature of the low energy, high energy transition hysteresis. The existence of this phenomenon remains a puzzle although it may be assumed that a polarization of a certain type may develop in the EIM a configurational or site association change which must be reversed by a high energy, counter polarization before the normal conductance transition may develop. The existence of any such secondary transition is obviously closely associated with the lipid interactions.

V. EFFECTS OF SPECIFIC IONS ON THE EIM MEMBRANE PROPERTIES

A. Calcium and Magnesium Ions: Steady State

As stated earlier, only small, twofold differences occur in EIM channel permeability to the cations, sodium, potassium, magnesium, and calcium, in the normal open state. In addition, it may be observed, as in Fig. 16, that with any of these ions as the sole cation used in preparation of the electrolyte, the normal conductance and switching parameters of the EIM membrane are not greatly changed.

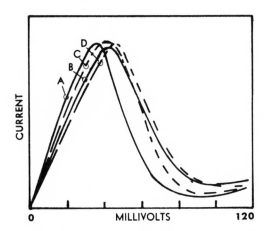

Fig. 16. Effect of cation on current-potential relations in the EIM membrane. Membranes were generated from phosphatidylethanolamine-tocopherol solutions in 0.1 M electrolyte and reacted with EIM as usual. Curves are normalized to equal peak heights from ramp potential recordings. A. Basic curve obtained with NaCl electrolyte. B. KCl electrolyte. C. $MgCl_2$ electrolyte. D. $CaCl_2$ electrolyte.

Despite the apparent equivalence of these ions when present as the sole cation, the divalent ions as minor constituents markedly alter the EIM membrane characteristics. Mueller and Rudin (14, 16) indicate that calcium in trace quantities is essential to the development of the EIM conductance and to the gating of the EIM channel. This seems analogous to the requirements of excitable membranes in cells, but this requirement has not been confirmed in our tests. With dialyzed EIM, extracted from Enterobacter cloacae, the introduction of EDTA into the electrolyte tended to inhibit the EIM reaction and change the position of the switching regions, but both reaction and gating could be observed even in 0.001 M EDTA with several different types of membranes.

When a divalent ion is introduced at low concentrations into a sodium chloride electrolyte on the same side of the membrane as the EIM, the EIM membrane parameters are altered. This is demonstrated in Fig. 17 which shows the I(V) curves for the membrane in 0.1 M NaCl (histidine buffer) and the curves subsequent to introduction of mangesium ion to a concentration of 0.005 M (ramp potentials). Several changes are immediately apparent in these curves. The central region increases in slope, indicating a large increase in open state conductance. In addition, the positive switching potential is increased, while the negative switching potential is decreased by a similar amount, as if the entire I(V) curve has been translated through the

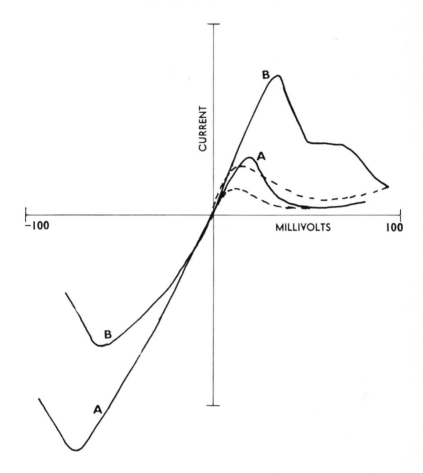

Fig. 17.   Effect of asymmetric low concentrations of a divalent cation in a univalent electrolyte.   The brain phospholipid membrane was generated in 0. 1 M NaCl and Curve A obtained after reaction with EIM.   MgCl$_2$, to 5 mM, was then added to the EIM side of the membrane and Curve B measured two minutes later.   Tracings of original recordings.

origin   toward the positive sectors.   Calcium causes exactly the same phenomena, differing only in that the potential shift may be about 20% greater while the increase in open state conductance may not be as great.

The potential shift is the most interesting of these phenomena, since it bears a striking resemblance to shifts induced by calcium ions in a number of excitable cells (63-68).   However, the change in open state conductance, which is the simplest to deal with, will be discussed first.

The increase in open state does not always develop. It may be almost entirely absent in certain lipid preparations which show a great willingness to interact with the EIM. Conversely, phospholipid preparations which have deteriorated through oxidation or hydrolysis may form membranes in which the calcium ion interaction may actually cause a decrease in open state conductance and a failure to show reversibility in recovery from the closed state. This latter behavior is evident even in membranes containing good lipids when heavier divalent ions or transition elements are substituted for calcium. Thus, barium, zinc, or aluminum ions may completely block the transition from closed to open state although not interfering with the switching off reaction.

When the conductance increase does occur, the change appears to be due to activation of latent EIM channels, or alteration of the equilibrium between open and closed states rather than an increase in the conductance of existing open channels. This is evident in the conductance jumps observed during the formation of EIM channels. These appear to remain perfectly constant in magnitude with or without a divalent ion. The divalent ion sometimes appears to improve the efficiency of the EIM reaction, increasing the rate of development of the conductance, but this could be due to an equilibrium shift favoring open states. It is unlikely that the divalent ion stimulated increase in the number of available channels is due to de novo adsorption and penetration of the excess EIM. This is precluded by the time factor, since the divalent ion may double, within less than a minute, the conductance of a membrane in which the base conductance required five to fifteen minutes to develop.

The observed potential shift of the I(V) curves is most strikingly comparable to the Gilbert and Ehrenstein descriptions of divalent ion induced shifts in the squid axon in high external potassium environments (63). These relations are strengthened by the quantitative analysis of the divalent ion effects. Figure 18 illustrates the shift in each of the switching potentials for two kinds of membranes as a function of magnesium ion concentrations. The brain phospholipid fraction used for the one set of results consits mainly of phosphatidylethanolamine with some phosphatidylcholine and other lesser substances. Despite a substantial difference in probable membrane site charge densities in the two types of membranes, curves for the degree of shift for all three switching potentials are essentially inseparable for both types of membranes.

The data of Fig. 18 is replotted in Fig. 19 on a semi-logarithmic basis together with additional data for calcium ion and for other types of membranes. With the exception of the low concentration points for magnesium ion, these curves resemble those which might be expected for an equilibrium association between the divalent ion and a site in the membrane,

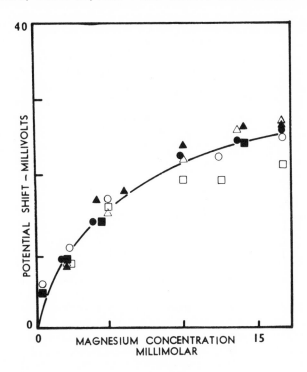

Fig. 18.   Relation of the switching potential shift to magnesium ion concentration.   Membranes were generated in 0.1 M NaCl.   After determining the switching potentials (dI/dV = 0) from the current-potential curves in absence of magnesium, magnesium chloride was added to the EIM side of the membrane at the indicated concentrations and the positions of the switching potentials redetermined two to five min later.   Each point represents averages for at least three membranes.   Open points are for brain phospholipid-cholesterol-sphingomyelin membranes; closed points for sphingomyelin-tocopherol membranes.   Triangles, potential shift for negative switching potentials, S1-; circles, shift for low energy positive switching potential, S1+; squares, shifts for high energy switching potential, S2+.

M + S = MS.   Such an association was proposed by Gilbert and Ehrenstein for the squid axon assuming that the divalent ion may neutralize the fixed site charge, thus creating a double layer potential to which the membrane channels are sensitive (63).

In the Gilbert and Ehrenstein treatment equilibrium curves corresponding to those of Fig. 19 should shift to the left or the right according to the

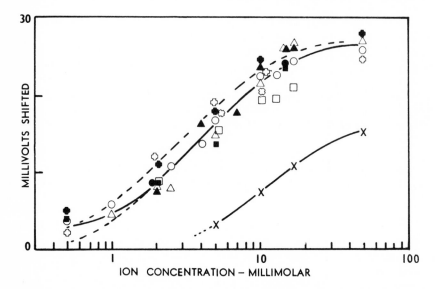

Fig. 19. Quantitative relation of switching potential shifts to divalent ion concentration and membrane composition. Procedures as in in Fig. 18. Open points, brain phospholipid-cholesterol-tocopherol membranes; closed points sphingomyelin-tocopherol membranes; triangles, circles and squares indicated S1-, S1+, and S2+, as in Fig. 18; X, S1+ for cholesterol-oxidized cholesterol-decane membranes; crosses, shifts for S1- and S1+ with $CaCl_2$ instead of $MgCl_2$.

strength of the association for the ion site interaction. The greater association strength should shift the curves toward the left. The amplitude and the asymptotic maximum shift of the curves should be determined by the density of the fixed sites contributing to the association within an appropriate range of the channel to affect the channel.

In the analysis of the present data, with reference to this model, the following points must be recognized. The calcium ion produces curves falling to the left of the magnesium curves in similar membrane systems but with amplitude appearing to remain unchanged. In absence of lipid fixed sites in the cholesterol membranes, the curves are shifted to the right with an apparent decrease in amplitude. The introduction of a charged lipid into the neutral cholesterol membrane reverses the rightward shift to bring the curves back in line with those for the phospholipid membranes and also to increase the amplitude. Finally, it must be realized that the data for these curves was collected from numerous membranes differing in conductance or density of EIM channels by orders of magnitude in many cases. This factor does not appear significantly to alter the equilibrium or amplitude.

Most of these data appear to favor, qualitatively, the double layer influence suggested by Gilbert and Ehrenstein. Thus, the difference between calcium and magnesium is consistent with the expectation that the association constant for calcium and the fixed sites should be greater than that for magnesium. Similarly, the decrease in amplitude in absence of lipid charges in the cholesterol membrane is predicted by the relation of amplitude to charge density in the interface. The rightward shift, due also to change in charge density, and its correction by insertion of a charged lipid, is somewhat anomalous, unless it may be assumed that the lipid charge density affects the equilibrium with the remaining interaction sites, which should be largely in the EIM or associated adsorbed proteins. This may be reasonable if it is assumed that some binding may include lipid-protein bridging.

The lack of response in potential shift or amplitude to the wide variation in EIM site density may have a number of explanations. It could indicate that channel sites are not themselves significantly involved in the binding phenomenon. The constant response could also be due to the distinct possibility that the EIM channels may exist largely in islands rather than distributed individually and randomly throughout the membrane surface. Within such active island areas the channel density might be rather uniform. In addition, the contaminating proteins of the EIM preparation may adsorb to the interface and provide a uniform effective charge distribution even in the absence of lipid charges.

Thus, the ion-fixed site association model appears reasonable, qualitatively, for the observed phenomena. However, the quantitative application of the Gilbert and Ehrenstein model may not be reasonable for these membranes. As Cole points out (69), a treatment based upon uniform distribution of charges in the effective vicinity of the channel may not be quantitatively applicable to widely spaced point charge concentrations. In these membranes, the channel density is about 10^4 to 10^7 channels per cm^2, giving a spacing of about 10^3 to $10^6 \mathring{A}$ between channels. These spacings are obviously large in relation to the area or depth of the channel. Consequently, it does not appear realistic to attempt quantitative evaluations based upon the uniform distribution model. This is reinforced by the deviation of the low concentration points, which consistently show a greater potential shift than should fit a calculated equilibrium curve. This suggests some significant interaction with a component far below the normal association levels and may be related to contaminants.

It must be emphasized that the effects discussed above depend on the divalent ion cations on the side treated with EIM. Although some similar shifts appear with transmembrane addition, these are very small in comparison with those discussed above.

B. Calcium and Magnesium: Kinetics

It is not entirely unexpected that the divalent ions should also introduce
changes in the switching kinetics of the EIM membrane since these are
normally affected by polarization amplitude in relation to the steady state
I(V) relations. As seen above, the effect of the divalent ions is to translate
these steady state curves through the axis toward more positive values. An
illustration of the alteration in transitional kinetics is given in Fig. 20.
The first set of curves shows the current responses to either positive or
negative potential pulses, with adequate rest in between each pulse to permit
recovery to the steady state, zero potential, resting conductance. Comparison of the positive pulses for magnesium-free and magnesium-plus
membranes shows little more than the change in open state conductance

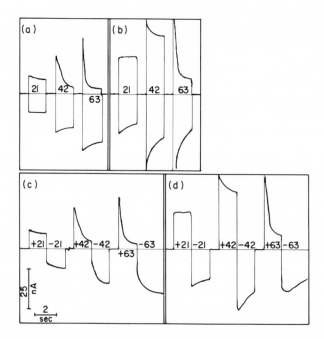

Fig. 20.   Change in pulse response kinetics in presence of divalent
ions. Brain phospholipid-cholesterol-tocopherol membranes. (a) Current
responses to increasing positive or negative potential pulses in absence of
divalent ions. (b) Same sequence with 5 mM $MgCl_2$ on EIM side of the
membrane. (c) Current responses to positive potential pulses followed
immediately by a negative potential pulse in absence of $MgCl_2$. (d) Positive-
negative pulse sequence response with 5 mM $MgCl_2$ on the EIM side of the
membrane.

(increased amplitude of the low potential pulses) and the potential shift, indicated by the requirement for increased potentials to induce the rapid resistance transitions. The negative pulses show similar relations, with the increasing currents on the negative pulses minus magnesium indicating only that the curves are highly asymmetric so that the maximum conductance region lies well into the negative sector and also showing a rather slow transition to open state. With magnesium added, this is altered so that the maximum conductance lies in the positive sector and the negative current responses show a decreasing current during the pulses. A more interesting aspect of changing kinetics is shown in the second series of curves, in which each positive pulse is immediately followed by a negative pulse. In the magnesium-free membrane, the return to maximum conductance in the negative pulse is rather slow, as observed in negative pulses above, with a time constant considerably greater than the time constant for the switching off response. However, after addition of the magnesium ion the negative pulse, following a transition to a positive closed state, is remarkably different. It now shows a rapid intial current increase to a maximum, followed by a slower decrease to the steady state value for that potential. This response is most interesting. Why should there be a bimodal response of this nature? It may be concluded that it suggests two important relations. First, the magnesium ion greatly accelerates the switching on from the positive off state and it appears to support very strongly the three-state kinetics discussed in a previous section. It seems apparent here that channels which have been switched off by a positive potential, upon being subjected to a negative potential large enough to switch off, must first pass through the open state (with negative current increase) before making the transition to the negative closed state. This behavior, however, would not have been detected except with the great acceleration in rate of switching on in the presence of magnesium ion. The rate of switching to open state in the magnesium-free membranes is considerably lower than the rate of switching to either of the closed states. If similar rates had been maintained after the magnesium-induced potential shift (which permits the negative switching region to become readily accessible), the initial rising-current phase of the negative pulses would be eliminated by the more rapid turning-off reaction. With the apparent great increase in the rate of switching on induced by magnesium, the full transitional sequence, positive closed $\rightleftarrows$ open $\rightleftarrows$ negative closed, becomes obvious.

Thus, it is apparent that the divalent ions have some effects on switching kinetics in addition to those which might be expected from a simple translation of the I(V) curves to alter driving potentials at any point along the curve. No mechanism for the increase in transition rate can be advanced at this point any more than for the increased development of open state channels.

C. Calcium and Magnesium: Permeability Control and Polarization

The potential shifting propensies of the divalent ions give rise to some intriguing speculations on their effects in cellur functions. It may be seen that the two-faced behavior of the divalent ions, referred to briefly in the introduction, may not be so surprising, and there should be a definite capability for control of cellular permeability in either direction depending upon the polarization of the cell.

A demonstration related to these possibilities is given in Fig. 21. A membrane with steady state characteristics in absence of magnesium or calcium represented by the curves in A was polarized at about +30 mV to clamp the membrane in the closed state indicated by the circled point. Then, recording the change in current with time at this polarization

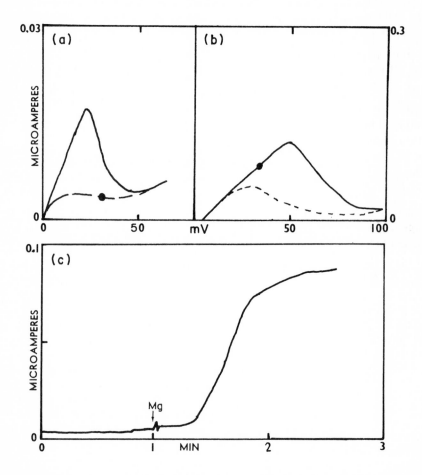

magnesium chloride, to 5 mM, was added to the EIM side of the membrane system. After a short induction period, the conductance rose rapidly to a new steady state level as much as 50 times greater than the original conductance of the polarized membrane. A reexamination of the steady state curves (ramp potentials) showed that the change was due to the shift in the switching region to a point beyond the polarizing potential so that the EIM channels were allowed to open.

Using an ion gradient to maintain polarization, with a suitable external circuit resistance, the same phenomenon may be translated into a membrane depolarization during the increasing current. This phenomenon may have interesting implications in cellular depolarization phenomena. Abood and Gabel (70) have proposed that cellular excitablity may be associated with reversible release of magnesium from ATP-enzyme sequestered states through hydrolysis of the ATP. The demonstration here that low concentrations of magnesium may create a depolarizing condition seems to support the logic of their arguments. It seems quite possible that magnesium ion sequestered in an ATP complex would have little effect until an appropriate stimulus activated the hydrolysis of ATP and release of magnesium. It may be indicated here that the positive polarization by external sources in the model experiment illustrated would correspond to the negative polarization of a normal cell, with the magnesium acting upon the external surface of the membrane causing, potentially, a potassium efflux.

D. Organic Cations

A number of cationic surfactants in the form of long chain alkylamines (e. g. , dodecylamine and octadecylamine) or quaternary ammonium ions (e. g. , dodecyl trimethyl ammonium bromide) will produce potential shifts similar to those found with the divalent ions. They have so far not been observed to cause the increase in open state conductance that is frequently observed with magnesium ion. The cursory quantitative measurements which have been made with such substances indicate that the primary amines are roughly as effective, on a molar charge basis, as the inorganic ions. Their activity tends to confirm the concept that the basis for the potential

---

Fig. 21. Divalent ion control of membrane conductance under polarizing conditions. Brain phospholipid-cholesterol-tocopherol membrane in 0. 1 M NaCl. The current-potential characteristics before addition of magnesium are indicated in (a). A polarizing potential of about 30 mV was used to clamp the membrane in a high resistance resting state, as indicated by the circle. Little change in current developed with time at the constant polarizing potential, as indicated in (b). Upon addition of $MgCl_2$ to 5mM current at constant potential increases rapidly to a new equlibrium value. After this change has developed, the current-potential relations had changed to those shown in (c).

shift is the alteration of the Donnan potential through introduction of positive charges into the interface by adsorption or association. A further examination of their influence under a variety of conditions, such as those
examined with the divalent ions, could probably settle some of the questions
left hanging in the divalent ion studies, such as the possibility that bridging
interactions by the inorganic ions may be of some importance. Bridging,
of course, could not be accomplished with the alkylamines and differences
should be observed under conditions where this was a factor.

However, some physiologically active amines produce a rather markedly
different effect. As pointed out earlier, analgesic amines produce apparent
diffusion potentials in their interactions with the basic lipid bilayer membranes. When these substances are introduced into the EIM membrane
systems, a different type of response is developed, as illustrated in Fig. 22.

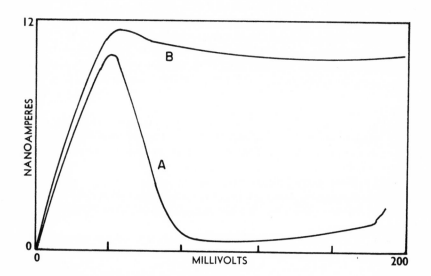

Fig. 22.   Procaine inhibition of switching off transition in the EIM
membrane. Brain phospholipid membranes were formed in 0.1 M NaCl.
Current-potential relations in absence of procain are indicated by Curve
A. After addition of 5 mM procaine (adjusted to pH 7), the current-
potential relations indicated by Curve B were observed. In contrast with
the divalent ions (Fig. 17), the procaine does not shift the switching potentials but either partially inhibits closing the EIM channel or provides a
separate leak channel. For these membranes, the effect shown is maximal
and changes only slightly with a ten-fold change in procaine concentration
in either direction.

In this case, where procaine is added to the EIM side of the membrane, instead of a shifting potential there is a large increase in the closed state conductance. On the other hand, the procaine added across the membrane (equivalent to being inside the cell) appears to have little or no effect upon the membrane characteristics at low concentrations except for development of a "diffusion potential" shift of the curves. It is apparent though, that with an effect of the same nature developed by the EIM side (outside of cell) addition of the procaine could lead to a cellular depolarization or greatly altered kinetics in an action potential.

Thus, it is apparent that specific amines may interact in an entirely different manner than the divalent ions. In this model the divalent ions had little or no antagonist effect upon the procaine, in contrast with the situation in cells. This may not be surprising however, considering the differences which probably exist in channel dimensions and probably in functional group structure.

Something of a disappointment was the behavior of acetylcholine with the EIM membrane. In view of the effects of the other substances, it was suspected that an ion such as acetylcholine might also act as an immobile ion, creating a Donnan potential alteration as a basis for depolarizing actions in synaptic transmission. Tests with acetylcholine proved negative, however, with little or no change induced in the I(V) characteristics. Nor did the acetylcholine have a significant effect upon the properties of membranes already altered by interaction with magnesium or calcium ions; nor did it act to inhibit the calcium ion potential shift. But, this information should not be taken to preclude an activity of this nature in the cellular systems. It is obvious that the relatively large effective aperture of the EIM channel differs greatly from the presumably tight channels in cellular membranes and this difference alone, without consideration for specificity of binding, might be sufficient to prevent an analogue action of acetylcholine in the EIM channel.

E.   Polycations

Biological polycationic molecules were found to create an effect very much like that of the divalent ions. This effect is best illustrated with protamine as in Fig. 23. This shows that protamine at extremely low concentrations creates a potential shift equivalent to those produced by the divalent ions at hundreds of times greater weight concentrations. This means that the protamine is more effective on a molar basis than the inorganic ions, even considering it on the molar ratio of the individual arginine groups, which comprise the cationic units of the protamine. Similar effects may be obtained with synthetic polyarginine and electric eel acetylcholinesterase at about the same weight concentrations as the

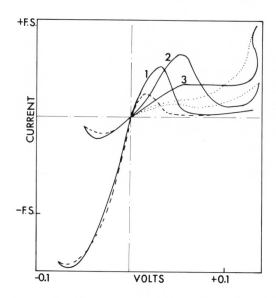

Fig. 23. Potential shift and conductance increase in the EIM membrane due to treatment with protamine. The initial characteristics of the membrane (brain phospholipid-cholesterol-tocopherol) are indicated by Curve 1 (full scale = ± 3 nanoamperes except at higher negative polarization where full scale = ± 15 nanoamperes). Ten minutes after addition of protamine to 0.5μ g/ml the characteristics, under potential control conditions, were changed to those of Curve 2 (full scale = + 100 nanoamperes). Under current control conditions, as in Curve 3, an N-shaped negative resistance appears with the protamine (full scale = ± 200 nanoamperes).

protamine, and with polylysine (synthetic) and cytochrome C, but at much higher concentrations.

Despite the seemingly high effectiveness of the protamine, it must be noted that the weight concentration of protamine exceeds the EIM by 10 to 100 times. This contrasts with the excess of 10^8 to 10^{10} for the divalent ions (molar basis) assuming a 100,000 molecular weight for EIM. Still, protamine may be at least 10^6 times as effective as the divalent ions in altering membrane polarizations.

These generalized cation-induced shifts of the switching potentials underline once more the probability that the effect is at least partly that of changing the membrane surface potential. However, in addition to showing an effect which could be attributed to a much greater association constant than that for the divalent ions, these polycations also produce a larger shift amplitude than the divalent ions. While the divalent ions appeared to have assymptotes in the area of 30 mV maximum shift, the protamine and acetylcholinesterase have produced shifts in excess of 50 mV, in some cases appearing to bring the negative switching region almost into the origin or across into the positive sectors. Further, the protamine and calcium effects are additive, or possibly synergistic would be a better term. The development of the protamine shift is slow relative to that induced by the divalent ions, but the introduction of magnesium along with or after the protamine leads to a very rapid shift with an amplitude as great as 80 to 100 mV.

With protamine and polyarginine, but not with the other polycations, another effect may be observed. This is the abrupt increase in conductance at higher potentials shown in the curves of Fig. 23. This sort of punch-through may also be induced in the EIM membrane with a number of detergent-type surfactants at very low concentrations and may also be induced by some EIM preparations, but generally not with the stability induced by protamine. Under current control conditions, the V(I) curves show an S-shaped negative resistance behavior in addition to the N-shaped curves found under potential control. This could be related to the punch-through mechanisms proposed by Coster (71) for the sudden increase in conductance of some algal cells under polarization. This is attributed to asymmetric channels, one side being cation selective, the other end being anion selective. It is apparent that there is a possibility for development of a channel of this nature in the EIM membrane where the normal channel is cation selective and could be altered by adsorption of protamine at one end to form an anionic mouth which could act in the manner proposed by Coster.

Mueller and Rudin have outlined further specific effects of protamine and polyarginine on the EIM membranes (14, 16, 72). It is possible, with proper titration and lipid composition, to develop action potentials and potential oscillations analogous to those found in several cellular functions. Both phenomena are attributed to the possibility that, under the appropriate ionic potential biasing and with the proper ionic conditions, a stimulus may cause the high resistance resting state to switch initially to low resistance with depolarization and then, through adsorption of and interaction with the protamine, the channel becomes anion selective, providing an overshoot and setting the stage for recovery and subsequent return to the normal resting state. This mechanism substitutes a cation to anion selectivity transition in place of the normal potassium-sodium interchange of mammalian cells but provides a basis for assumption of a single channel operation in the action potential.

VI. CONCLUSIONS

The preceding sections have presented a rather broad outline of diverse properties of the lipid bilayer membrane and its EIM-modified form, the excitable lipid bilayer membrane, to give some indication of the mechanisms by which external agents, particularly drugs and ions, may affect membrane function. While it may not be possible to extrapolate directly from the behavior of these synthetic models to the cellular membrane, it is reasonable to assume that a number of the observations could have some validity in reference to the action of drugs on cells. The salient points of membrane action, and their implications in cellular function, may be reviewed at this point.

The lipid bilayer membrane has been rather widely accepted as a reasonable model of the basic core of some cellular membranes, although it is certainly not universally acclaimed as the only membrane structure. In its basic form, however, its electrical properties have no relation to those of the excitable cellular membrane. Treatment with a specific protein, EIM, initiates development of a steady state negative resistance characteristic, closely resembling that which may be observed by voltage clamp measurements in cells such as the squid axon.

In the synthetic membranes it now appears reasonably certain that the variable conductance created by EIM results from penetration of a complex protein into the membrane to form transmembrane channels or pores. These channels may exist either in open or closed states with the transition between open and closed state obeying simple equilibrium kinetics but having a transmembrane voltage dependency superimposed upon the normal equilibrium factors.

The combined chemical and electrical potential dependency may subject the on and off equilibrium of the EIM channels to control by a large number of chemical and physical factors. Divalent ions and ionizing drugs, which may interact with sites in the vicinity of the channels, should be able to alter the equilibrium status of the channels through their effects upon the interface potential. More direct effects may also be involved, with substances which may interact with some portion of the channel to alter the energy barriers in the switching phenomenon.

This seemingly simple mechanism can have an extremely wide latitude in operational characteristics, since the membrane composition with respect to proteins and lipids will have both direct and indirect effects upon the channel configurations and upon the interactions of specific chemicals. Versatility may be increased by the existence of two switching regions, positive and negative, with asymmetric characteristics and opposing responses to chemical agents. Thus, it is possible to produce either an increase or decrease in membrane conductance with a single substance acting upon one side of the membrane simply as a function of the polarity of a polarizing potential across the membrane.

As a biological model the synthetic membrane has several implications. The development of ion-selective porous structures in a lipid bilayer matrix provides a precedent for assuming the existence of related structures in vivo.

It may be easy to imagine, with this as background, that some cellular membranes can consist of a mosaic of protein units penetrating a lipid matrix. Specific transport functions could be assigned to different protein units while the density of protein channels would reflect the degree of activity required for any given membrane function.

On the other hand, the observation here that the open state conductance of the EIM channel is almost independent of the lipids surrounding it may also permit a broader interpretation. In the synthetic model the lipids are apparently essential only as insulators and modifying agents and, perhaps, as interchannel binders. This may imply that the cellular membrane can be functionally developed without a lipid bilayer core. It seems reasonable to assume that a protein structure related to the EIM would still be able to function in a similar manner, even if inserted into a protein matrix, with a minimal structural complement of lipids. Transitional kinetics should be maintained in either structural model. For an excitability function, the membrane must be capable of undergoing large changes in conductance so that parallel leak conductances must be minimized. The required insulating surroundings for the channels in this case would probably be most economically provided by a lipid bilayer structure with minimal protein penetration. Thus, it appears reasonable to assume that the basic lipid bilayer structure might constitute a substantial proportion of the membranes of electrically active cells or organelles, while membranes oriented toward other transport functions may be largely constituted of protein with minimal bilayer structure.

The studies with the lipid bilayer membranes appear to place a substantial base under the suggestions of some investigators that the excitable cell owes its properties to transitions of a channel organization between two stable states. It is apparent from the information presented that the two-state transition may exist in a protein-based channel and that the transition may be affected by many of the chemical and drug factors which alter the excitability functions of cells.

ACKNOWLEDGMENTS

This research was supported by 6570th Aerospace Medical Research Laboratories, Aerospace Medical Division, Air Force Systems Command, Wright Patterson Air Force Base, Ohio, under Contract AF 33(615)-5240, and by the Army Research Office, under Contract DA 49-092-ARO-50. We are indebted to Drs. Paul Mueller and Donald O. Rudin for their frequent advice and discussions about the lipid bilayer membranes, and to Dr. Robert E. Kay for criticisms of this manuscript.

REFERENCES

1. W. R. Loewenstein, Ann., N.Y. Acad. Sci., 137, 441 (1966).

2. W. J. Adelman, Jr. and R. E. Taylor, J. Gen. Physiol., 45, 93 (1961).

3. S. Hagiwara, K. Naka, and S. Chichibu, Science, 143, 1446 (1964).

4. G. Abood and R. W. Gerard, J. Cellular Comp. Physiol., 43, 379 (1954).

5. D. E. Goldman and M. P. Blaustein, Ann., N.Y. Acad. Sci., 137, 967 (1966).

6. D. O. Shah and J. G. Schulman, J. Lipid Res., 6, 341 (1965).

7. H. L. Rosano, H. Schiff, and J. H. Schulman, J. Phys. Chem., 66, 1928 (1962).

8. I. R. Miller and M. Blank, J. Colloid Interface Sci., 26, 34 (1968).

9. E. Rojas and J. M. Tobias, Biochim, Biophys. Acta, 94, 394 (1965).

10. C. van Breeman and D. van Breeman, Biochim. Biophys. Acta, 162, 114 (1968).

11. A. M. Monnier, A. Monnier, H. Goudeay, and A. M. Rebuffel-Reynier, J. Cellular Comp. Physiol., 66, 147 (1965).

12. P. Mueller, D. O. Rudin, H. T. Tien, and W. C. Wescott, Circulation, 26, 1167 (1962).

13. P. Mueller, D. O. Rudin, H. T. Tien, and W. C. Wescott, Nature, 194, 979 (1962).

14. P. Mueller and D. O. Rudin, J. Theo. Biol., 4, 268 (1963).

15. P. Mueller, D. O. Rudin, H. T. Tien, and W. C. Wescott, in Progress in Surface Science (J. F. Danielli, K. G. Pankhurst, and A. C. Riddiford, eds.), Vol. 1 Academic, N.Y., 1964, pp. 379-393.

16. P. Mueller and D. O. Rudin, J. Theo, Biol., 18, 222 (1968).

17. H. T. Tien and A. L. Diana, Chem. Phys. Lipids, 2, 55 (1968).

18. J. Taylor and D. A. Haydon, Discussions Farady Soc., 42, 51 (1966).

19. H. T. Tien and A. L. Diana, J. Colloid Interface Sci., 24, 287 (1967).

20. L. L. M. van Deenen, Ann., N.Y. Acad. Sci., 137, 717 (1966).

21. H. Davson and J. F. Danielli, The Permeability of Natural Membranes, Cambridge Univ., 1942.

22. C. Huang and T. E. Thompson, J. Mol. Biol., 13, 183 (1965).

23. H. T. Tien and E. A. Dawidowicz, J. Colloid Interface Sci., 22, 438 (1966).

24. T. E. Thompson and C. Huang, J. Mol. Biol., 16, 576 (1966).

25. R. L. Cherry and D. J. Chapman, J. Mol. Biol., 40, 19 (1969).

26. T. Hanai, D. A. Haydon, and J. Taylor, Proc. Roy. Soc. (London), Ser. A., 281, 377 (1964).

27. T. Hanai, D. A. Haydon, and J. Taylor, J. Gen. Physiol., 48, 59 (1965).

28. C. D'Agostino, Jr. and L. Smith, Jr., in Biophysics and Cybernetic Systems (M. Maxfield, A. Callahan, and L. S. Fogel, eds.), Spartan, Washington, D.C., 1965, pp. 11-23.

29. F. A. Henn, G. L. Decker, J. W. Greenawalt, and T. E. Thompson, J. Mol. Biol, 24, 51 (1967).

30. C. Huang, L. Wheeldon, and T. E. Thompson, J. Mol. Biol, 8, 148 (1964).

31. H. T. Tien, J. Phys. Chem., 71 3395 (1967).

32. D. A. Haydon and J. L. Taylor, Nature, 217, 739 (1968).

33. R. C. Bean and W. C. Shepherd, 150th Meeting, American Chemical Society, Atlantic City, N.J., June 1965.

34. S. Noguchi and S. Koga, J. Gen. Appl. Microbiol., 15, 41 (1969).

35. T. E. Andreoli, J. A. Bangham, and D. C. Tosteson, J. Gen. Physiol., 50, 1729 (1967).

36. V. K. Miyamoto and T. E. Thompson, J. Coll. Interface Sci., 25, 16 (1967).

37. R. Pagano and T. E. Thompson, J. Mol. Biol., 38, 41 (1968).

38. A. Petkau, Biophys. J., 9, A-34 (1969).

39. R. C. Bean and H. Chan, in The Molecular Basis of Membrane Function (D. C. Tosteson, ed.) Prentice-Hall, Englewood Cliffs, New Jersey, 1969, pp. 133-146.

40. D. Papahadjopoulos and J. C. Watkins, Biochim. Biophys. Acta, 135 639 (1967).

41. W. R. Redwood and D. A. Haydon, J. Theo. Biol., 22, 1 (1969).

42. S. Ohki and A. Goldup, Nature, 217, 458 (1968).

43. F. A. Henn and T. E. Thompson, Ann. Rev. Biochem., 38, 241 (1969).

44. P. Lauger, W. Lesslauer, E. Marti, and J. Richter, Biochim, Biophys. Acta, 135, 20 (1967).

45. B. Rosenberg and H. C. Pant, Biophys. J., 9, A-31 (1969).

46. D. Paphadjopoulos and S. Ohki, Science, 164, 1075 (1969).

47. R. C. Bean, W. C. Shepherd, and H. Chan, J. Gen Physiol., 52, 495 (1968).

48. A. L. Diana and H. T. Tien, Biophys. J.. 8, A-25 (1968).

49. W. D. Seufert, Nature, 207, 174 (1965).

50. J. Bielawski, T. E. Thompson, and A. L. Lehninger, Biochem. Biophys. Res. Commun., 24, 948 (1966).

51. U. Hopfer, A. L. Lehninger, and T. E. Thompson, Proc. Natl. Acad, Sci. U.S., 50, 484 (1968).
52. V. P, Skulachev, A. A. Sharaf, and E. A. Liberman, Nature, 216. 719 (1967).
53. E. A. Liberman and V. P. Topaly, Biochim, Biophys, Acta, 163, 125 (1968).
54. L. D. Kushnir, Biochim. Biophys, Acta, 150, 285 (1968).
55. R. C. Bean, W. C. Shepherd, H. Chan, and J. T. Eichner, J. Gen. Physiol.,53, 741 (1969).
56. G. Ehrenstein, H. Lecar, and R. Nossal, J. Gen. Physical., 55, 119 (1970).
57. W. D. Stein, Nature, 218, 570 (1968).
58. W. Drost-Hansen and Anitra Thorhaug, Nature, 215, 507 (1967).
59. I. Tasaki and I. Singer, J. Cellular Comp. Physiol., 66, 137 (1965).
60. D. Nachmansohn, Proc. Natl. Acad. Sci. (U.S.), 61, 1034 (1968).
61. H. M. Fishman and R. I. Macey, Biophys. J., 9, 140 (1969).
62. H. M. Fishman and K. C. Cole, Federation Proc., 28, 234 (1969).
63. D. I. Gilbert and G. Ehrenstein, Biophys. J., 9, 447 (1969).
64. B. Frankenhauser, J. Physiol., 137. 245 (1957).
65. B. Frankenhauser and A. L. Hodgkin, J. Physiol., 137, 218 (1957).
66. B. Hille, J. Gen. Physiol., 51, 221 (1968).
67. F. J. Julian, J. W. Moore, and D. E. Goldman, J. Gen. Physiol., 45, 1217 (1962).
68. S. Hagiwara and K. Takahashi, J. Gen. Physiol., 51, 221 (1968).
69. K. C. Cole, Biophys, J., 9, 465 (1969).
70. L. G. Abood and N. W. Gable, Perspectives Biol. Med., 1 (1965).
71. H. G. L. Coster, Biophys. J., 5, 670 (1965).
72. P. Mueller and D. O. Rudin, Nature, 213, 603 (1967).
73. U. Hopfer, A. L. Lehninger, and W. J. Lennarz, J. Membrane Biol., 3, 142, 1970.
74. S. G. A. McLaughlin, G. Szabo, G. Eisenman, and S. M. Ciani, Proc. Natl. Acad. Sci., U.S., 67, 1268, 1970.
75. R. C. Bean, Fed. Proc., 30, 1282Abs, 1971.

DATE DUE

ILL (PAM)			
595 1531			
MAY 9 1996			
ILL (PAM)			
595 1531			
due 6/6/96			